DISCARDED

The Acquisition & Divestiture of Petroleum Property

THE ACQUISITION & DIVESTITURE OF PETROLEUM PROPERTY

A GUIDE TO THE STRATEGIES, PROCESSES AND TACTICS USED BY SUCCESSFUL COMPANIES

JIM HAAG

Disclaimer
Two years ago this book began as an idea to simply capture some of the "best practices" that were being used by companies who were successful at buying and selling oilfield property. The text grew as my personal experiences with coworkers and industry professionals were added to the observations which had been made over a long period of time. The book represents a synthesis of these experiences and does not necessarily reflect the processes used by any one particular company.

Copyright ®2005 by PennWell Corporation
1421 South Sheridan Road
Tulsa Oklahoma 74112
800 752 9764
sales@pennwell.com
www.pennwell-store.com
www.pennwell.com

Cover design by Ken Wood
Book design by Mark Brown

Library of Congress Cataloging-in-Publication Data
Haag, Jim, 1934 -
The acquisition and divestiture of petroleum property: a guide
    to the strategies, processes and tactics used by successful
    companies
Includes biographical references and index.
ISBN 1-59370-045-8
1. Petroleum-land grabs-strategy. I. title.
HD9685.U6 M4758 2004
333.793'217'0975--dc21

All rights reserved. No part of this book may be reproduced, stored in a retrieval system, or transcribed in any form or by any means, electronic or mechanical including photocopying or recording, without the prior permission of the publisher.

Printed in the United States of America.

1 2 3 4 5   09 08 07 06 05

Contents

Preface ix

Acknowledgements xiii

1 **Successful Company Profiles** 1
   The Players 1

2 **Motivating Factors** 11
   For the Seller 11
   For the Buyer 13

3 **Non-motivating Factors** 17
   For the Seller 17
   For the Buyer 19

4 **Seller Evaluation** 23
   Fiscal Decisions 23
   Retention Value 26
   Tax Consequences 27
   Earnings Write-Downs 29
   Financial Impact 31

5 **Marketing Options** 33
   Auctions 34
   Brokers 42
   In-House to Industry 44
   Solicit Co-Owners 45

6 **The Divestment Process** 47
   Activity Timeline 47
   Data Room Requirements 49
   Third-Party Reserves Report 51

7 **The Acquisition Process** 55
   Steps to the Process 55
   Proactive and Reactive Approaches 57
   Data Room Visit 59
   Reserves Assessments 60
   Financing Options 61
   Product Price Considerations 62

8 Unique Life Cycle Risks 65
Exploration 65
Delineation 66
Development 66
Production 67
Redevelopment 68
Sunset Period 69

9 Valuation Methodologies 71
Fair Market Value 71
Seller Versus Buyer Viewpoints 73
Market Multiples—Asset Acquisitions 74
Market Multiples—Historical Figures 79
Operational Concerns 81

10 Determining the Price 83
Impact of Various Factors 83
Multiple Field Packages 85
Sensitivity Analysis 86
Product Price 88
Reserves Growth 90

11 Like-Kind Exchanges 91

12 New Field Discoveries 93
Characteristics 93
Reasons for Sale 94
Buyer's Evaluation 95
Deepwater Example 96

13 Constructing the Offer 99
Options and Alternatives 99
Appraised Value 101
Strategy Amount 104
Sales Package Examples 106

14 Preferential Rights 113

15 Bonding Protection 117

16 Negotiating the Agreement 121
Terms and Conditions 121
Effective Date 123

17 Due Diligence 125

18	Government Approvals	129
19	Corporate Sales Programs	131
20	Industry Activity	135
	Historical Acquisition Prices	135
	Justification of Elevated Multiples	139
21	Company Mergers	141
22	Accountability	147
23	Lessons Learned	151
	Processes	151
	Interdisciplinary Team	152
	Evaluations	152
	Negotiations	153
	Execution	155
24	Case Histories	157
	Gulf of Mexico: Exploration Well Purchase/Sale (Category 1 Deal)	158
	South Louisiana: Acquisition Proposal for Old Inland Field (Category 1 Deal)	160
	South and North Louisiana: Purchase/Sale for a Nonoperated Package (Category 1 Deal)	163
	Gulf of Mexico: Purchase/Sale of Sunset Properties (Category 1 Deal)	166
	South Louisiana: Purchase/Sale With Proactive Co-Owner (Category 1 Deal)	167
	East Texas: Purchase/Sale of Property With Engineering Uncertainty (Category 1 Deal)	169
	Gulf of Mexico: Exploitation Strategy Purchase/Sale (Category 1 Deal)	171
	Gulf of Mexico: Gas Price Impact on Assignment of Interest (Category 2 Deal)	173
	Gulf of Mexico: Tax Impact From Assignment of Interest (Category 2 Deal)	174
	Gulf of Mexico: Trade of Property, Evaluation Expertise (Category 3 Deal)	176
	South Louisiana: Innovative Approach to Property Trade (Category 3 Deal)	177
	Gulf of Mexico: Trade of Property for Consolidating Efficiencies (Category 3 Deal)	179

Preface

This guide is intended to be a helpful tool for anyone who wants to understand the process of buying and selling producing oilfield assets. The decision to buy or sell a property, the evaluation methodology that is employed, the strategy used in the construction and delivery of an offer, and the aspects of effectively negotiating and closing a transaction involving the exchange of title to a producing property are abstract activities. It is an inexact science that is approached in a unique fashion by each company. Learning the concepts that are presented in this guide by the trial and error method can be terribly costly, and has indeed resulted in bankruptcy or forced merger for many companies.

Huge profits and losses occur throughout the industry as a result of the purchase and sale of petroleum property. The learning curve can be quite expensive and time consuming for buyers and sellers if the parties are not well versed in every phase of the process.

All buyers are cognizant of the *winner's* curse, which is the unintended result of having the high bid for a competitively marketed asset, only to realize later that the price paid for the acquisition is so high that the transaction is not profitable. Inexperienced or unskilled acquisition teams who overpay due to excitement or flawed analysis can only hope that their company is large enough to withstand the error and that it is not repeated.

Most domestic producing fields are bought and sold many times before depletion. In the past 25 years, more than 6,000 domestic transactions valued in excess of $650 billion were announced publicly. It is obvious from these figures that huge sums of money are placed at risk regularly in acquisition and divestiture activity. Companies involved in this business must be experts in the principles that are presented in this guide to be successful.

Active companies optimize their portfolios by buying and selling property to meet business objectives. The expertise that is needed is invaluable and maintained as a core competency by those who are most successful. The ability of the team to merge geological, geophysical, engineering, land, legal, marketing and financial data into a seamless analysis is just the first step. The information is then risked appropriately within what is generally a very tight time frame. Management presentations and bid strategy follows rapidly, culminating in the submittal of an offer. It is very gratifying to the participants when this high-energy effort concludes with a successful transaction.

This guide addresses the acquisition and divestiture of individual producing or discovered non-producing assets or packages of such assets. Pure exploration deals involving promoted wildcat drilling or primary term-lease trades are not reviewed since the probabilistic evaluation of exploration assets is unlike the more deterministic evaluation of producing assets. Corporate-level transactions are not addressed either, although many of the concepts presented herein are the building blocks for the analysis of the producing assets of a corporate transaction. The myriad of complicated financial and legal factors that are investigated for corporate transactions beyond asset value are not discussed either, as the topics are beyond the scope of this text.

Each company has a preferred economic model. The majors commonly use in-house programs that have evolved through generations of users. Most independents utilize one of the many popular commercially available programs that are occasionally upgraded by the software provider. Some of the smallest operators are still comfortable with 'back of the envelope' calculations that have served them well in the past.

Regardless of the evaluation methodology that is chosen, any technique can be used to analyze the purchase or sale of producing property when applied in a consistent manner for both the acquisition effort and the post-acquisition accountability review.

There are many excellent references that can guide a company in the evaluation of producing property. Taken collectively, the resources provide several perspectives and approaches to analyze value, uncertainty, and risk. Economic evaluation and decision analysis textbooks, articles in industry magazines, monographs, short courses and society presentations each contribute to the body of knowledge that exists to improve the accuracy of the evaluations that are done in support of the acquisition and divestiture of oilfield property.

Acknowledgments

References are made to the following organizations in the text to recognize their contributions to the industry, specifically their efforts to organize and educate the professionals who are engaged in property evaluations:

The Society of Petroleum Evaluation Engineers (SPEE) – The SPEE was formed in 1962 to answer a long-standing need for a professional organization which would bring together, for their mutual benefit, specialists in the evaluation of petroleum and natural gas properties. The organization has authored several useful publications for the evaluations engineer. The recently published monograph entitled *Perspectives on the Fair Market Value of Oil and Gas Interests* is a collection of fundamental evaluation principles and example problems that can be the basis for any company's fair market value (FMV) estimation procedures. The annual *Survey of Economic Parameters Used in Property Evaluation* is an informative compilation of the forecasts of evaluation parameters by a cross section of industry professionals.

The Council of Petroleum Accountants Societies, Inc. (COPAS) – COPAS was created in 1961 to provide a forum for discussing and solving the more difficult problems related to accounting for oil and gas. Their publication entitled *Property Acquisition Checklist* (1995) is an excellent tool that can be used to ensure that all aspects of the accounting side of the transactional process are reviewed.

The author expresses his appreciation to the following companies for permission to include information and exhibits as follows:

> The Scotia Group, Inc.: Scotia is a full-service oil and gas advisory firm founded in 1981 with extensive worldwide experience. Services provided include expert opinion reports, acquisition evaluation, exploration analysis, reservoir studies and simulation, research and technology applications, strategic planning and risk analysis, and reserves analysis and property valuation. Scotia provided access to their transactional database, which contains data from more than 6,500 domestic transactions beginning in 1979, as well as plots showing unit acquisition costs from 1982 to the present.
>
> The Oil & Gas Asset Clearinghouse: *The Clearinghouse,* a Petroleum Place Company, is a full service acquisition and divestiture consulting firm facilitating the exchange of oil and gas assets. They offer a comprehensive array of evaluation, technical, and marketing services to meet the specific needs of oil and gas industry clients. They are the industry leader in petroleum property auctions and provided all of the historical data regarding the growth of auctions and the prices paid at auction that are referenced in this text.

The author also wishes to acknowledge his family members who provided much of the fuel to fire this effort. My wife, Annette, has given me support and encouragement throughout my career and was my ever present cheerleader during the book writing process. My sons, Jeremy and Chris, and my daughter-in-law, April, continue to inspire me by their daily commitment to tackle difficult challenges in their studies and research. My parents also deserve special credit, as they instilled in our family the value of lifelong learning, a virtue that gives us the energy and excitement to pursue self-improvement.

There were many hardworking professionals and caring mentors at Texaco, Inc. who I was fortunate to spend a great deal of time with on more projects than I can count. These good people along with other industry friends who I worked with making evaluations and closing transactions made every day a good day at the office.

One individual who I must single out and thank for his encouragement is Gene 'Skip' Wiggins. Skip is a good friend and the consummate professional. He was the first to see the vision of my class notes being taken to a higher level—without him there would be no book.

Last but not least, I owe Marla M. Patterson, Editorial Team Leader at PennWell Books, my sincere thanks for accepting this manuscript for publication and guiding the effort to a successful conclusion.

SUCCESSFUL COMPANY PROFILES 1

The Players

Buyers

Successful buyers have the ability to create value, sometimes very significant value, above the purchase price. This can be done in a number of ways:

- The purchase price is simply much less than the actual field value.
- The product price rises above the analysis premise, creating unexpectedly higher profit margin in cash flow.
- The buyer has access to previously unused technology or develops new technology.
- The buyer executes a successful exploitation or exploration drilling program.
- The buyer is able to develop lower confidence reserves or accelerate long life reserves.
- The buyer's operating costs are lower than the seller's operating costs.
- The buyer is able to abandon a property cheaper than the seller.

Achieving the first point requires that the seller make a mistake. The second point contains an element of luck. Neither of these situations (relying on a mistake or luck) is a reliable basis for making an acquisition, although at times both have been known to happen unexpectedly for the buyer. Either of these situations should result in a successful purchase.

The remaining points have a common theme—in the pre-acquisition analysis the buyer proactively identifies the potential for more cash flow than the seller creates in the retention scenario analysis. There are companies who have the technical expertise to do an outstanding job of pre-acquisition analysis, to identify upside potential and to operate very efficiently. They rarely make a mistake, do not stray far geographically beyond their niche area of expertise, and are focused on quality rather than quantity. The potential savings by being able to lower field operating costs is shown in Figure 1–1. Operator 1 is evaluating the purchase of Field Area B from Operator 2.

A hallmark of all buyers is that they have a strong desire to acquire volume. The natural depletion of the resource base causes

	Field Area A Operator 1	Field Area B Operator 2
Well count	30	5
Producing rate	4500 bpd	500 bpd*
Lifting cost	$4/bbl	$5/bbl

bbl = barrel; bpd = barrels per day.

*Assume a 7-year rate life in Field Area B at 500 bpd = ~ 1.3 million barrels reserves.

Operator 1 has an operating cost that is $1/bbl less than operator 2.

The cost savings for Operator 1 compared to Operator 2 for the life of Field Area B because of operating area expansion = $1.3 million.

Operator 1 can pay up to $1.3 million above Operator 2's retention value, and still make a profit on the purchase. Overhead cost savings would also be captured to increase the profitability of the purchase.

Fig. 1–1 Savings from operating cost reduction

most companies who are not successful explorers to be on the prowl for deals on a continuous basis. Many companies do not want to play the costly, high-risk exploration game, and as a result must make acquisitions to grow or even to just maintain company size.

There are many properties that are divested that were not being marketed by the seller while the buyer's analysis was in progress. It was only due to the aggressive nature of the buyer that a deal was made. Because it is generally difficult to get the owner of a high quality property to consider a sale, buyers have become very imaginative in their proposed transaction structures. Terms that share the risk and reward of future development or exploration are relatively common. At times high risk upside potential is recognized by the seller, and the seller chooses not to currently fund the development of the potential. In this case, the buyer may need to share some of the upside with the seller as a carrot to get the seller to divest the property. This benefits the buyer, the seller and the landowner, because the buyer has the intention to develop reserves that may have gone undeveloped.

Price protection is also a measure that limits the exposure of both the buyer and seller from extreme fluctuations in wellhead price. It can be incorporated into the contract as shown in Figure 1–2. A seller may want a share of the additional revenue for a few years after closing if prices rise substantially above an agreed upon forecast to guard against a significant increase in cash flow to the buyer. A buyer may want the same protection after closing to guard against a drop in product price. In the latter scenario, if the product price drops below a mutually accepted forecast, the seller would return a portion of the purchase price to the buyer.

A buyer can also lock in a relatively predictable stream of cash flow by hedging the near term price of the liquids and/or gas production. To increase the chances that the acquisition will be profitable, the production volumes and cost projections in a purchase analysis must be relatively accurate for the first two years of the evaluation. To guard against a price drop that would cause an otherwise successful acquisition to lose money, the buyer can lock in

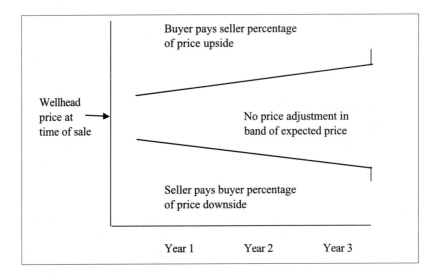

Fig. 1–2 Price protection in sales contract

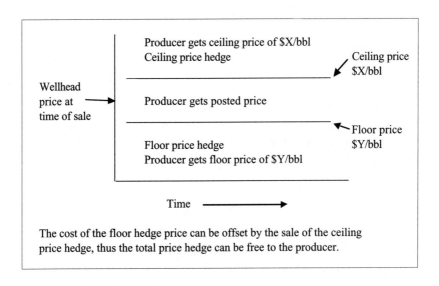

Fig. 1–3 Price protection by hedging

the profitability of the deal by hedging (or paying a fee to guarantee a floor price for the production) as shown in the simplified Figure 1-3. Some hedge arrangements or *collars* allow participation in crude oil or natural gas price increases while others will cause a sacrifice of price upside.

A successful buyer makes prudent, selective acquisitions, rarely overpays for any property, and, in the aggregate, has an ongoing program that provides an acceptable rate of return. If a company finds itself being the high bidder on most of the offerings it bids, it is a sign of paying too much relative to the market. Some of the acquisitions under this scenario would likely not reach payout and have unacceptable economics.

Sellers

The most successful sellers are also accomplished buyers. They are experts in property evaluation and know the nuances and pitfalls of the transactional process.

Historically, the majority of the sellers through the early 1990s were the major oil companies, and the majority of the buyers were the smaller independents. Figure 1–4 is an example of the relative portfolio sizes of each type of company for a representative size transaction. Both buyer and seller would upgrade their respective portfolios with transactions of this nature.

Beginning in 1998, several majors merged while a number of the independents did the same. Domestically, this created a new three-tier hierarchy of company sizes: the huge majors, the large independents, and the small independents. This caused a change to the divestment activity of the past. Each new post-merger company was driven to pare its portfolios of the smallest fields and those that did not fit the new company strategy. As a result, the majors sold large packages of properties to the large independents, and the large independents sold smaller packages to the small independents. Figure 1–5 shows the newer brand of field sales with the three sizes of companies that are now involved in the domestic marketplace.

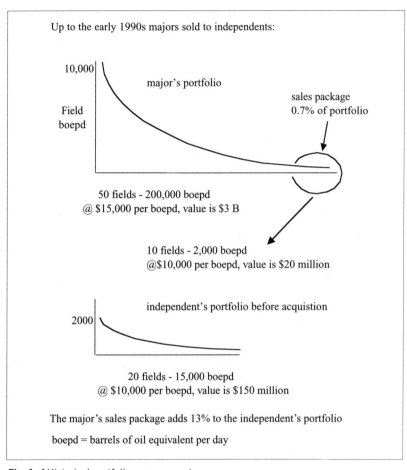

Fig. 1–4 Historical portfolio management

As suggested previously, the major integrated oil companies that once owned most of the fields in the United States have divested huge portions of their portfolios during the past decade. Many of the majors own a mere fraction of the domestic properties that were their core producers in 1990. Each company has developed a variety of processes for its sales programs. The approaches to the sales vary and are based on staffing requirements and field characteristics, including size, value, geography and associated abandonment or environmental liabilities.

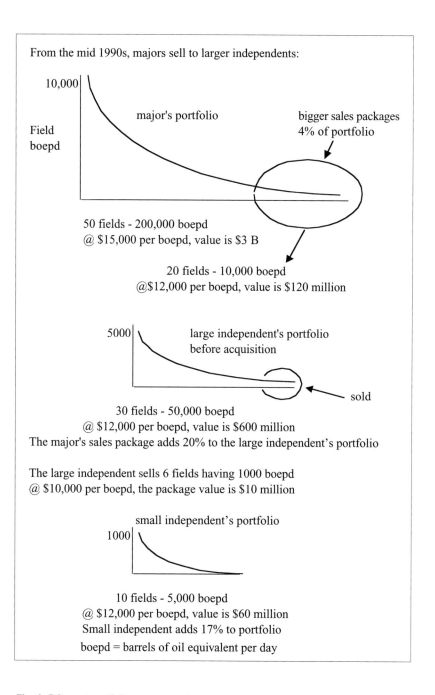

Fig. 1–5 Current portfolio management

Independents have purchased the majority of these divested properties. The buyers have created significant value by engaging in well work that has increased daily production above the pre-sale decline curves. There are many reasons that smaller companies have found and developed additional value that was not exploited by the majors:

- Highly competent technical staffs have fewer distractions and can focus on new acquisitions.
- Monetary incentives are common for those employees who successfully add value.
- There is a willingness to exploit smaller targets that have challenging economics.
- Operating practices may not be encumbered by corporate policies that drive up costs.
- Economic hurdle rates for capital expenditures are lower for independents.
- Opportunities abound in fields that were categorized as *non-core* by a major, since those fields do not receive daily attention.

The following illustration demonstrates that if a small percentage of a field's resource base is not valued at the time of sale by the operator, it can provide a significant profit opportunity for the purchaser (Fig. 1–6).

The majors are not paid at the time of sale for the unrisked future revitalized value of a property. However, the independents take significant risk to pursue this upside value, and, on a risk-adjusted basis, the majors do receive adequate compensation. The characteristics noted previously indicate that the smaller companies simply have a more aggressive business model compared to the majors. The sellers who are most successful know who the most aggressive buyers are and will contact them preferentially when selling a property. There is a need and a place in the oil industry for both players.

The ultimate measure of success for a seller is to close a transaction for a price that includes a premium above the retention value of a property. To achieve this, upside potential is presented to the marketplace on an unrisked basis in a credible fashion. Since the calculation of reserves or upside is interpretive and buyers risk assets differently, an aggressive yet plausible interpretation of the data generally results in the seller obtaining a good offer.

Field is operated by a major—ultimate reserve 10 MMbbl.

Major produces 8 MMbbl and then decides to sell.

Independent purchases field at $6/bbl, for $12 million.

Upside of 5% of ultimate reserve is attributed to lower-hurdle-rate opportunities.

Impact of the 500,000 bbl of upside to independent (5% of 10 MMbbl):

Gross revenue @ $20/bbl	$10,000,000
Investment @ $2/bbl	–1,000,000
Lifting cost @ $4/bbl	–2,000,000
Taxes @ $3/bbl	–1,500,000
Royalty @ $3/bbl	–1,500,000
Net revenue @ $8/bbl	$ 4,000,000

bbl = barrel; MM = million.

The $4 million profit from producing the low hurdle rate reserves will recover one third of the purchase cost!

Fig. 1–6 Reserve additions from lower hurdle rates

Relationships

The previous sections of this chapter discuss the bricks and mortar aspects of successful company efforts. As is the case in most transactions, the intangible benefits of strong business relationships also come into play. Most companies have preferred partners with a past history of negotiating fairly and paying market prices.

When a major has closed a number of transactions with a smaller independent successfully, it is natural for the major to want to negotiate a private sale with the independent for future activity rather than going through the time and effort of hosting a data room to industry. This is exactly the position and reputation that a small company works tirelessly to achieve. A few purchases via this route can build an attractive portfolio for a small or start-up independent company. Many of the smaller but successful companies in the Gulf of Mexico have built their businesses impressively during the 1990s with this approach.

Small companies frequently have employees who know a major's portfolio based on prior employment. These employees are aware of drilling leads that the major was not enamored with because of the perceived low reserves potential or high risk. Farmouts may be made or joint ventures will be formed to get the wells drilled, frequently leading the way to a property sale later between the parties. At times, the mature production will be sold while the major retains an overriding royalty interest in upside potential. In this way, if the drilling results are significant the seller has managed to retain some revenue stream from the production. In the Gulf Coast area, many smaller companies have built sizable portfolios using this approach.

In most cases, the smaller companies that are successful have built a niche that suits their personnel, risk tolerance, capital depth, and geographic expertise. Making acquisitions outside their niche expertise is risky if done too rapidly or without proper preparation. Conversely, leaving the safety of their niche too late could limit their prospects for growth.

MOTIVATING FACTORS 2

For the Seller

There are many reasons that motivate a company to sell an asset. In some cases, one reason is enough; in others, it is a collection of factors that combine to justify the sell decision. The reasons frequently cited are:

- The geographic location is no longer attractive.
- The asset has poor financial metrics (high lifting cost or low profit margin).
- The property to be sold is no longer a good fit in the company portfolio.
- The business environment has changed (increased regulatory scrutiny or landowner issues).
- The scheduled capital expenditures lack the desired profitability.
- The upside potential of the property is now lacking (no remaining drilling).
- Existing production is risky (wells may have mechanical problems).
- Reserves recovery is in jeopardy (performance indicators have worsened).

In addition to these sell incentives, which are generated internally, a common reason to sell is the receipt of an unsolicited offer that exceeds the retention value. At times the operator of the asset cannot understand how an offer can be so high and will react by selling the field to maximize its present value.

The progression of a geographic area from being a profitable portfolio of early life field discoveries to one of primarily a sunset-period liability is shown in Figure 2–1. The field characteristics change considerably during the life cycle from discovery to depletion.

The annual decline rate of the production complex was approximately 10% for the 40-year period from 1960 to 2000.

1960	1980	2000
1,000,000 bpd	125,000 bpd	15,000 bpd
10 fields	8 fields	5 fields
Huge exploration effort	Limited exploration	Exploration farmed out
High investment rate	Modest exploitation	Loss of acreage
Robust construction activity	Abandonment cost peaks	Environmental oblig peaks
Volume growth	High cash flow	Operations costly
		Idle wellbores dominate
		No exploitation remaining
		TIME TO SELL!

bpd = barrels per day.

Fig. 2–1 Loss of core area—south Louisiana wetlands

The progression of field characteristics through several decades of rising and falling activity provides a representative account of the life of a major producing area. The dominant operating parameters of drilling and production in the first half of the cycle are gradually replaced by high operating costs, loss of leasehold because of landowner demands, fewer upside opportunities, and high abandonment costs.

If a company does not want to shoulder the abandonment and environmental remediation costs of a huge operating area, an exit strategy should be planned while the value of the production and attractiveness of the upside potential are still high enough to entice a financially qualified buyer.

For the Buyer

There are many reasons that motivate a company to purchase an asset. A common thread among buyers is that they will have a compelling or strategic reason to want a particular property. Acquisitions are very competitive; without the passion to make the deal, a prospective buyer will probably not be successful.

It has been said that for every five solicitations that pass across a manager's desk, one fits the company strategy. For every two that are evaluated, a bid may be made on one. For every five bids made, one may be successful. These data suggest that one deal will result from 50 solicitations. Obviously, these odds can be very discouraging and show that:

- It is costly for a company to engage in the acquisition process.
- It is more costly to engage in the acquisition process if the buyer has limited acquisition experience.
- It is difficult to recover from overpaying for an acquisition.

A company may decide to play the acquisition game for many reasons; the most common include:

- The company can fulfill a desire to enter a promising geographic area and establish production immediately.
- The company has confidence that it can increase production and create value beyond the seller's vision.
- The company can take advantage of an opportunity to exploit technical expertise or apply technological advances.
- The company can maintain its critical mass by countering the natural depletion in other areas.
- The company can distribute its overhead across an increased asset volume and thus improve financial metrics.
- The company may have had an unprofitable exploration program for a prolonged period.
- The company may be responding to Wall Street analysts, who consider volume growth to be the *flavor of the day.*

These are the typical drivers for companies that participate in the acquisitions process. Atypical drivers that skew the logical purchase price upward and push a bid into the *outlier* range include the following:

- A company wants to go public, and an acquisition is the only way to reach the desired market capitalization quickly to proceed with an initial public offering (IPO).
- There may be substantial operational savings as a result of natural synergies.
- The product price forecast is significantly beyond the competition.

Bids that are based on these factors are very tough to beat and usually leave the losers shaking their heads and wishing they had known the competition better. It is worth the effort to keep abreast of the recent prices paid by active buyers in the area—this information can lead a company to redeploy the internal resources to a different acquisition that would have a better chance of success. An important discussion early in the acquisition process concerns who the other bidders may be, what is driving their buy attempts, and how aggressive they might be.

Clearly, entry into the acquisition arena is not to be taken lightly. It requires specific skill sets, a well-defined strategy, and the fortitude and resources to play the game frequently. To avoid gambler's ruin—which is the potential to go broke (or have an unsuccessful acquisition program) if too few acquisitions are made must be a key component of a company's objectives.

NON-MOTIVATING FACTORS   3

For the Seller

A standard practice in the industry is for companies to study the assets of other firms that are located near their existing operating areas for potential acquisition. This work is generally done privately with public data. When the analysis concludes with the company wanting to acquire the asset, an unsolicited offer is sent to the owner of the asset. Upon receipt and evaluation of the offer, the owner will make a decision to reject it, to counter, or to negotiate toward closing with the potential purchaser. There are many reasons why the property's owner may not want to sell.

One common factor that influences the owner to hold the property, even if the offer provides a fair return on retention value, is that it is extremely difficult to replace production. Whether the reserves would be assumed to be replaced either by exploration or by acquisition activity, the results of pursuing either method are very uncertain. If the owner of the property has no need for the cash and the property is a good strategic fit in the portfolio, the offer would need to be exceptional for most producers to sell. Reserve replacement is a large issue that would not be overlooked.

Another key consideration with the divestment of a producing asset is the impact on the remaining properties in the portfolio. If other fields operate in conjunction with the field that is the subject of the offer, the loss of operating synergy could easily cause the lifting costs of the other fields to rise. This could reduce the profit margin of the offset production to a level that may not be tolerable.

A frequent disincentive for management to review an offer for a large field, particularly one with upside potential, is the commitment of resources that is required to do an appropriate analysis. The current situation in most companies is that staff is occupied with ongoing projects, and the time is not available to also engage in a field study. An associated problem occurs when management wants such an offer to be reviewed in detail, but only partially commits the necessary staff to the job. This normally results in a flawed analysis.

Compounding the difficulty of the decision is the potential for product price appreciation. As shown in Figure 3–1, a small price increase creates a disproportionately large improvement in profit margin, and a seller is generally not willing to give up this option value without some premium in the asset's sale price. The illustration shows that for a price increase of 50% per barrel, the profit will jump by 94% per barrel. This is obviously a huge margin improvement that would be welcomed by any purchaser.

Another disincentive to selling a producing property is the federal tax that is assessed on transactions when gains are realized. Typically, the sale price exceeds the remaining tax basis (capital expenditures that have not yet been written off as tax deductions). This capital gain creates a tax cost that the seller must pay. Thus, the seller needs a price that exceeds the value of the asset as well as the transaction's tax cost to break even on a cash basis. If the seller wants the buyer to pay for the tax cost in addition to paying a premium for the property's reserves value, it is unlikely the parties will negotiate to a price that is mutually acceptable, and a sale will not occur.

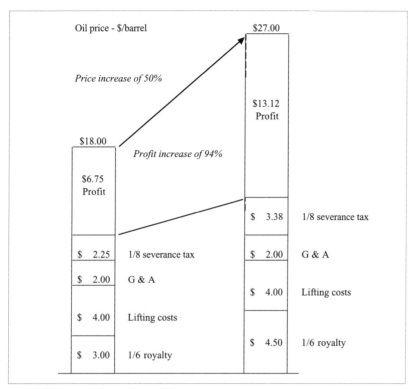

Fig. 3–1 Value gain from price uplift

For the Buyer

A number of factors can cause a buyer to back off from a potential acquisition after the first contact is made. Most buyers do their best to avoid this situation in order to maintain a good reputation and to conserve resources. However, when working with only public information a potential buyer never knows all the facts regarding a property. The reason that a buyer retreats is generally tied to one of the following:

1. **Indemnities required by seller:** Depending on the seller's position on granting various indemnities to the buyer, there can be a gap in this issue that is so wide that the sale will not be closed. When majors sell a field to a smaller company, it is quite common that the opening position is that no indemnifications will be granted to the buyer. The buyer must negotiate to gain indemnification in the purchase and sale agreement. If the offer is high enough to offset the loss of protection that is desired by the seller, the seller may soften its position and lessen its requirements to allow the sale to close.

2. **Discovery of undisclosed liabilities:** During the due-diligence process, environmental liabilities or deteriorated field or offshore platform conditions may be discovered that are so bad that the buyer does not want the responsibility to remedy the situation. If no amount of reduction to the purchase price gives the buyer comfort, the company will walk away from the deal.

3. **Financing questions or problems:** Many negotiations are begun with the knowledge that the buyer requires financing to close the deal. As the lending institution goes through its review process, there are many steps that can cause the lender to advise the buyer that the money will not be made available for the transaction. The reserve report must support the necessary reserve volume, and the cash flow analysis must show that there is a high likelihood that the funds will be repaid per the terms of the loan. Other associated factors (operator quality, buyer's financial status, pricing expectations, etc.) must also satisfy the lender's criteria. At times the value to be generated from the asset will require more capital than the organic cash flow from the property itself, creating another tier of investment capital that the buyer may require from outside sources.

4. **Lack of information:** Owners of properties may not have made public some transactions that impair their asset picture in order to achieve specific goals. For instance, deep drilling rights, exploration potential, or overriding royalty interests may have been carved out of the lease ownership to facilitate upside development. These actions dilute the revenue interest, value, and attractiveness of a property. Deals of this nature are rarely noted in the public domain and are only discovered when the company must divulge this information during the due-diligence process.

   The production performance of an asset may be worse than suspected. The lack of important data may be a result of a time lag in production reporting or the lack of pressure or field data being available. In other cases, a key well or facility may have a mechanical problem that may not be repairable. The buyer can reduce his offer to reflect the new information or may choose to disengage from the deal entirely.

5. **Delay requested by the seller:** In some cases the seller may be willing to divest the property for strategic reasons, but the tactical loss of volume from the sale would be unwise to report to Wall Street because analysts may criticize the divestiture. The seller may ask the buyer to wait until the end of the fiscal year or until some other positive event for the company occurs that would offset or replace the reserves lost.

   A delay of the transaction is generally not attractive to the buyer, as the resources are available for purchasing and working the asset now, and they may not be available later. Also, as the saying goes, "time is money," and the buyer would want to initiate the intended well work and make operational improvements now rather than later.

SELLER EVALUATION   4

Fiscal Decisions

Nearly every producing field is a sales candidate at some point in time. The reasons for selling and the associated timing vary with the circumstances as noted in chapter 2, Motivating Factors. In addition to those strategic reasons, company fiscal policy may dictate that the following economic factors govern the decision above all others:

- When money is needed for investments or debt service requirements, it may be necessary to monetize producing assets. The least desirable fields in the portfolio that can generate the necessary funds would be selected for sale in this situation. At times, the need is so great that a high-quality field must be sold. In these cases, a high market multiple would be expected.

- There are many fields that are worth more to another company than they are to the current owner. An example of this is when a company operates a limited portion of a much larger field, and the larger operator has a much lower unit cost structure. In this case, the smaller operator may solicit an offer from the larger operator, knowing the larger company may pay more than the smaller operator could ever realize in value by continued ownership.

- Some companies do not want to fund abandonment operations or environmental cleanup because of the uncertainty of the activity or because this type of expenditure is unbudgeted. The value of a producing property in its sunset years will eventually approach zero because of the pending abandonment obligation, even while the daily cash flow is positive. If the company does not want to be the party to perform the abandonment, then the field must be divested. Ideally, the field will be marketed while the reserve value offsets the future costs and still provides enough of a profit to attract a large pool of potential buyers.

Table 4–1 gives an easy spreadsheet for calculating future field value for use in predicting when fields should be sold. The annual cash flows are entered into the second column from the current year until time of abandonment. The spreadsheet calculates the present value of each year sequentially until depletion for any discount rate. The field's present value is totaled each year, showing that this example field should be marketed in 2008 at the latest, the year in which the present value is approximately equal to that year's cash flow.

The data can then be plotted as shown in Figure 4–1. The plot shows when the value loss begins to plummet as the influence of the abandonment cost is felt.

Figure 4–2 gives a generic illustration of the plot illustrated in Figure 4–1; it shows the decline in market value and annual cash flow over time for any producing property. When the property value of the field still exceeds one year of cash flow, it is the ideal time to market it, because the seller has produced the bulk of field value, but enough value remains that quality buyers would still have an interest in acquiring the field. Once the field value falls to a year's cash flow or less, the pool of quality buyers shrinks considerably. If the marketing effort is delayed to when the field value is negative when the end-of-life costs exceed the reserve value, a field is nearly impossible to divest by itself.

Table 4-1 Determination of present value in future

	Annual undiscounted cashflow		Future discounted annual cashflows						Discount factors
		2005	2006	2007	2008	2009	2010	2011	@ 10.5%
	$M	$M	$M	$M	$M	$M	$M	$M	
2005	1650	1571							0.952
2006	1250	1076	1190						0.861
2007	1550	1211	1335	1476					0.781
2008	1100	776	859	947	1047				0.705
2009	650	415	458	508	560	619			0.638
2010	250	145	160	176	195	215	238		0.578
2011	-1400	-732	-809	-893	-987	-1093	-1205	-1333	0.523
Present value in future:		4460	3192	2213	815	-259	-967	-1333	

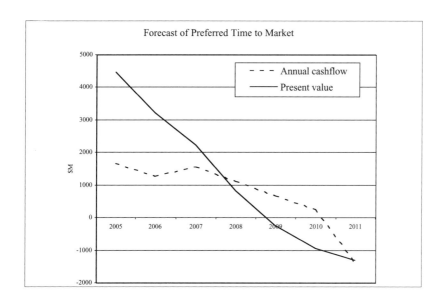

Fig. 4-1 Forecast of preferred time to market

Seller Evaluation

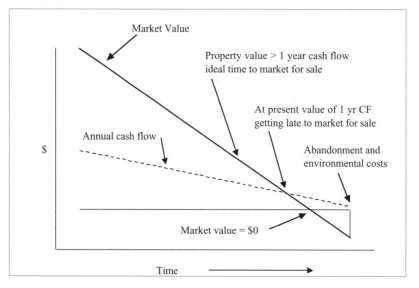

Fig. 4–2 Generic selection of marketing time frame

Retention Value

Once a field has been deemed a "sales candidate," one of the first deliverables is a calculation of its retention value. This is defined as the value of the property, managed as if it remains in the portfolio. The last phrase, "managed as if it remains in the portfolio," is important and will differ from one company to the next. Each company has its own internal metrics that serve as a guide, resulting in potentially different values by co-owners for the same field.

If the company is an aggressive exploiter of property, then undrilled opportunities should be fully valued in the analysis. If the company notes these opportunities but will not fund them, regardless of the reason, then a discounted value similar to farm-out value more accurately represents the holding value of those opportunities.

The holding value should be determined before the marketing effort begins because

- The reserve volumes and associated risks are anchored.
- The capital expenditure (CAPEX) opportunities, which will be in the *base case funding*, are chosen.
- Agreement for the value from all team members and management is obtained.
- Agreement is reached as to how much the sale price premium above the holding value should be.
- The expected tax and earnings impacts for the sale can be calculated.
- Ideas on why the property should not be sold are stated early in the process.

Tax Consequences

The seller's evaluation of tax consequences is determined using the field's retention value and an estimate of the selling price. The calculation includes a number of factors, including:

- The *field retention value*, which is defined as the after-tax net present value of the property
- The *tax cost*, which is defined as the tax paid to government entities on the taxable gain (i.e., the difference between the offer and the tax basis of the property)
- The *net proceeds*, which is defined as the offer amount minus the tax cost (if any) of the sale

Two examples of the tax calculation are shown in Figures 4–3A and 4–3B. These problems are used to determine if an offer for a property results in a profit after the transaction tax is paid by the owner of the property. The examples use the corporate tax rates that were in effect when most of the transactions discussed in this text were evaluated, that is before 2002, at which time the rates were lowered. The use of the higher rates that were in effect before 2002 for these examples gives a more vivid illustration of the tax impact at that time.

If a field has an after-tax retention value of $1 million and a tax basis of $800,000, and an offer is received for $1.3 million, does the analysis show a profit to make the sale? The company has an effective 35% tax rate.

The *tax cost* is the offer less the tax basis multiplied by the tax rate:
($1,300,000 – $800,000) x 35% = $175,000

The *net proceeds* is the offer less the *tax cost*:
($1,300,000 – $175,000) = $1,125,000

The margin of profit is the *net proceeds* minus the *field retention value*:
($1,125,000 – $1,000,000) = $125,000

In this example the seller makes a profit of $125,000 to sell the field. Because of the uncertainty of reserves determination and product pricing, a 12.5% premium over retention value would probably not be enough to create a compelling sell argument. The owner of the property would want to negotiate and increase the offer.

Fig. 4–3A Simplified tax calculation for a sale, example 1

If a field has an after tax retention value of $1 million, a tax basis of $200,000, and an offer is received for $1.3 million, does the analysis show a profit to make the sale? The company has an effective 35% tax rate.

The *tax cost* is the offer less the tax basis multiplied by the tax rate:
($1,300,000 – $200,000) x 35% = $385,000

The *net proceeds* is the offer less the *tax cost*:
($1,300,000 – $385,000) = $915,000

The margin of profit is the *net proceeds* minus the *field retention value*:
($915,000 – $1,000,000) = –$85,000

In this example the seller loses $85,000 if the sale is made. In cases such as this, where the buyer is already offering a premium and the seller has a low tax basis, a deal may not result even if both companies want the transaction to occur. A buyer will rarely have acceptable economics if the tax cost is a large percentage of the seller's price requirement, and the buyer is required to cover this cost.

Fig. 4–3B Simplified tax calculation for a sale, example 2

What the preceding discussion shows is that if the property has a low tax basis that results in a high tax cost to the seller, the buyer needs to assign much more value to the asset than the seller for a sale to occur. This difference in perception can be on the cost side or in the reserves, production rate, product price, or upside potential. Also, because the seller knows the asset better than the buyer, a prospective buyer needs an exceptional technical staff that can study an asset and find this additional value if it is present.

Despite the conclusion that a sale should not occur because of a loss in after-tax value, some properties are divested for strategic reasons even when the tax cost is not totally covered by the sales price.

Earnings Write-Downs

Earnings reported quarterly by publicly owned companies greatly influence the stock price of the company, and for this reason earnings losses or write-downs are managed or avoided whenever possible. A sale results in an earnings loss when the sales price is less than the remaining book value of the property. This situation is common when fields with high abandonment costs (including wells, facilities, or platforms) or environmental remediation (including pits, groundwater contamination, or the property surrounding wells or facilities) are sold during their sunset years.

The total incremental investment that was made in a property is summed to become the cumulative book value for the asset. As reserves are produced, a proportional share of the investment is written off as a cost in the earnings calculation. Thus, if the reserves are not overbooked or underbooked when the last barrel is produced at depletion, the last dollar of investment is written off with it.

In many fields, however, the reserves are overbooked. As fields are studied year after year, pressure to replace or increase reserves frequently results in booking higher risk volumes for sidetracks into attic reservoirs, drilling operations for smaller reservoirs, mechanically difficult plug down workovers to capture bypassed reserves, and performance reserves that require lower and lower reservoir pressures for recovery. These reserves do not comprise a large percentage of the field reserves when field depletion is many years away. However, closer to depletion, if the wells are junked or the reserves are deemed to be high risk or economically marginal to develop, not all companies would write off the reserves on a timely basis. A write-off would cause an increase to the unit of production (UOP) rate and lower the field earnings per barrel of oil equivalent (boe) produced. Thus, if a field sale occurs while the reserves are overbooked and the UOP rate has previously been kept artificially low, it is likely that an earnings write-down will occur.

Note: A sale occurs at three reserves levels for comparison.

Reserves value = $6/boe
UOP writedown rate = $3/boe
Abandonment cost = $3 million
Remaining investment starts at 2 MMboe reserves = $6 million

(For simplicity, the following figures are shown before tax impact and discounting.)

Reserves (MMboe)	2	1	0
Reserves value @$6/boe ($MM)	12	6	0
Abandonment cost ($MM)	3	3	3
Market Value ($MM)	9	3	−3
Booked Investment ($MM)	6	3	0
Earnings impact of sale ($MM)	+3	0	−3

boe = barrels of oil equivalent
MM = million
UOP = unit of production.

The figures show that the field should be sold before the reserves volume depletes to 1 million boe to avoid an earnings loss.

Fig. 4–4 Earnings impact at various sales points

With the passage of the Sarbanes–Oxley Act of 2002, these practices of allowing uneconomic or nonexistent reserves to remain on the books should be curbed substantially. The Act's penalties for misleading stockholders in publicly traded companies should deter fraudulent financial reporting by most corporations. Improved audit procedures and management reviews have also helped to ensure that the reserves that are reported as proved are determined according to Securities and Exchange Commission (SEC) guidelines.

If there were no abandonment costs and the booked reserves actually equaled the ultimate recovery, then there would rarely be earnings write-downs from asset sales. But this is not typically the case as all oilfields have end-of-life costs. At the time that field value is zero, the discounted value of the remaining reserves is equal to and offsets the discounted cost of abandonment. The investment that remains to be written off the books in these sale situations may be substantial. Figure 4–4 shows that the earnings impact for a field sale will vary depending on the timing of the sale relative to the remaining reserves.

Financial Impact

The prior discussions do not completely address the nuances of tax and earnings calculations; individuals more knowledgeable about financial matters must be consulted to provide this information using updated tax guidelines and the specific tax and earnings criteria that is appropriate. The prior discussions were meant to show that when a field is chosen as a sales candidate, one can easily make the following two determinations:

1. Whether or not the tax cost of a sale is likely to be covered by the expected sales price, and if not, how much the net tax cost may be
2. Whether or not the earnings impact of a sale is likely to be negative, and how bad it may be

The value of doing these two simple calculations before any effort is put into field evaluations or marketing efforts is obvious. If company management is not willing to absorb the anticipated tax cost or negative earnings impacts from a sale, the time-consuming effort and valuable industry goodwill is saved by deciding to hold the field before marketing even begins.

If management still wants to sell assets that will generate a negative financial impact (yet does not want the negative impact), one way to accomplish the objective is to package the fields with other fields that have high value and positive offsetting financial characteristics. Then, when the impact is calculated for the package, the transaction is not negative, but neutral. At times this requires making the difficult decision to sell assets that would not have been sold otherwise.

MARKETING OPTIONS 5

Every asset has unique attributes, and when the asset is coupled with the objectives and resources of the seller it becomes clearer as to what divestment strategy is the best match for the sale. Many options exist and each will have its advantages and disadvantages. Even after a sale has taken place, it is not always clear if the best method was picked. Thus, to be assured that an asset sale has the best chance of attracting a quality buyer and a high price, a decision-making process that considers all the available information should be used.

The basic building block for all of the marketing options is the creation of a data package that honestly and fairly represents the value of the property. The data must be gathered by the seller and massaged into a logical format that enables the buyers to quickly determine the proven value, visualize the upside potential, and understand the other characteristics of the asset. If this task is accomplished effectively, the sale should attract a wide audience resulting in a price at market value.

Auctions

Properties having low values are frequently sold at auction for several reasons:

- Producers generally prefer to assign their resources to activities that have a higher priority and generate a higher profit than in-house management of a low-value sales program.

- Statistically there is a relatively low risk that the high bid will be less than the seller's calculated retention value.

- If the seller auctions a number of properties at one time, there is a good probability that the combined total of the sales prices will exceed the total actual value of the properties.

- The companies that auction producing properties are very skilled and have a streamlined, effective process.

- Brokers who normally market larger fields on a commission basis generally don't market lower value property because their compensation would be too low. It can take as much effort to market a low-value asset as it does a high-value asset. (Commissions that are paid for low-value assets at an auction are higher percentages than typical broker's commissions.)

Prior to the year 2000, properties valued at more than $100,000 were not auctioned frequently. Sellers did not want to take the chance that a property would be lost to a low bid that may be substantially below the asking price. However, lots exceeding even a million dollars are sold now, indicating the success and acceptance of the auction process for higher value assets. The breadth of the audience in attendance has become reliable enough that the potential for the high bid being far below retention value is minimal.

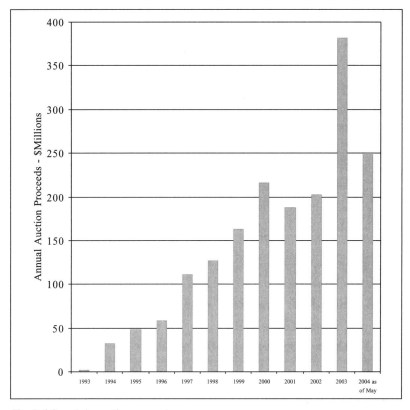

Fig. 5–1 Growth in auction proceeds

The Oil & Gas Asset Clearinghouse (O&GAC)—which provided the information that was used to prepare the exhibits in this section—handles approximately 90% of the petroleum property auction business in the United States. The organization's data clearly show that using auctions to sell producing property has become very popular.

With sales beginning in 1993, Figure 5–1 shows the impressive growth in auction volume, topping out in the year 2000 at more than $200 million in sales, which was the most active year in auction history up to that point. The 2001 and 2002 results trailed those of 2000, with the most often cited reasons being:

- Properties were much more profitable than expected because of the high product price, thus lower valued assets were still worth keeping.
- Producers did not need the cash because all of their assets were generating more cash flow than was forecasted.
- A sale of property would have increased the company tax obligation, which was probably already higher than expected.

In 2002, despite the continued increases in product price, the inventory of property that would have been taken to auction in the previous two years in the absence of a high product price was soon brought to market, and the auction volume in 2003 surged to nearly $400 million.

Tracking this impressive trend is the average sale size shown in Figure 5–2. The average live auction proceeds climbed from approximately $2 million in 1993 to nearly $50 million in 2003.

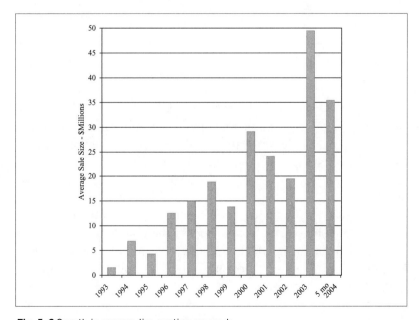

Fig. 5–2 Growth in average live auction proceeds

At times properties or wells are bundled for sale and sold in lots. The most common reasons for this are operational synergy and geographic proximity. Figure 5–3 shows the increase in the number of lots valued at more than $250,000 that were sold from 1993 (1 lot) to 2003 (nearly 450 lots). The average lot price for all lots sold has increased from $8,000 to $180,000 during the same time period, as shown in Figure 5–4.

It is obvious that the marketplace has given its approval to the auction process, and, based on the industry participation shown on these exhibits, the valuation of properties by buyers and sellers at auction has increased.

The statistics from the recent years after the onset of the sale of larger and more valuable assets need to be used cautiously if a seller wants to estimate what price multiple his assets may bring at auction. The sales price and price-per-barrel-produced figures begin to reflect the larger deals exclusively as they swallow up the statistics from the

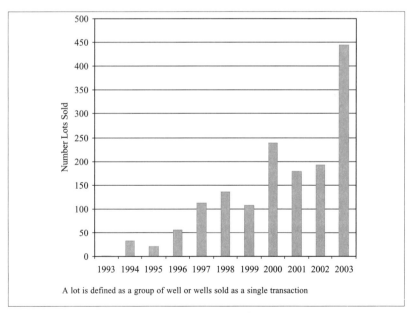

Fig. 5–3 Growth in auction lots valued at more than $250,000

Marketing Options

smaller transactions, thus it is unknown what the deal parameters are for the smaller transactions that at one time comprised the whole of the auction business.

One option for a seller is to elect to use a minimum bid figure, which serves as insurance that a property will not be sold below the lowest sale price that would satisfy the seller. The seller provides this minimum price to the auctioneer, who will start the bidding at a price that is slightly higher than the minimum. If no buyers participate, the auctioneer can drop down to the minimum, but not below. Properties that are assigned a minimum bid are occasionally not sold.

The closing documents for a property to be sold in this manner are made available for inspection by the seller prior to the event. Because the bidders must accept the closing documents as written, the lengthy negotiating process associated with other types of sales during which the buyers and sellers agree to the terms is eliminated. For this reason, all bidders are encouraged to review the closing documents before the auction and to adjust their bids accordingly if any terms or conditions create, in their opinion, a reduction to the asset value.

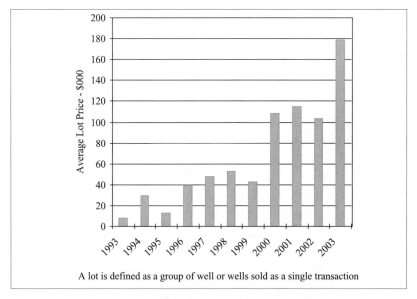

A lot is defined as a group of well or wells sold as a single transaction

Fig 5–4 Growth in average auction lot price

A distinction has been noted at times in the sales price multiples between property that is operated by a major versus an independent producer. Bidders appear to pay more per barrel for fields that are operated by a major for two reasons:

1. Majors are perceived to develop a given field to a lesser degree, at times to the point of neglect, compared to if a smaller firm had operated it. This is thought to be because the majors have a higher profitability threshold requirement for drilling and well work activity. A field operated by an independent is frequently thought to be *picked over* to the extent that few upside opportunities remain at the time of sale.

2. Majors are also perceived to maintain their fields in better condition than independents because of the attention given by the increased governmental and environmental scrutiny, internal controls guiding company activity, and generally having company personnel rather than contract personnel working the fields.

As a result of the above observations, at times the majors may obtain a higher price multiple in the auction process than the independents would receive for the same field.

Figure 5–5 shows production multiples that were received from May 2000 to February 2004 for working interest and royalty interest properties. The royalty interests obtained a significant premium over the working interests, capturing more than double the value per barrel produced. The working interest multiple increased from $10,000/barrels of oil equivalent per day (boepd) to $30,000/boepd, and the royalty interest multiple increased from $25,000/boepd to $75,000/boepd during the time period.

This escalation in value may appear to be unreasonable, but there is a strong correlation in value between oil price and the production market multiple, as is shown in Figure 5–6. This correlation is stronger than that which is shown when the oil price is compared to the market multiples of all U.S. transactions, probably for two reasons:

1. The time for a property to be identified as a sales candidate to the time when the transaction is closed is much shorter for the auction process than any other sales option.
2. There is very little nonasset value included in the properties that are sold at auction.

The data show that the production multiples trend high to the broader U.S. market. Given that the data are burdened by the sale of a significant number of low-value wells and fields, it is another

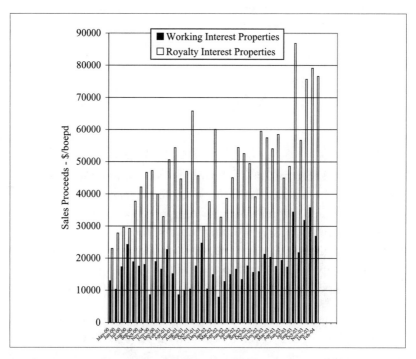

Fig. 5–5 Auction sales proceeds—comparison of working and royalty interest properties

indicator that there are many higher value properties sold at auction. The high average multiples show that the auction process has been very successful at drawing an attractive pool of buyers.

The high production market multiples obtained at auctions should make bidders extra cautious. Wells that may be forecasted to experience an accelerated decline, perhaps because of the onset of water production, should get a lower production multiple than a well producing on a more gradual decline, if it is producing by a pressure depletion drive mechanism. If a bidder gets swept up in bidding statistically at auction and acquires lots that have an accelerated decline in the near future, the profitability of these acquisitions will be very disappointing.

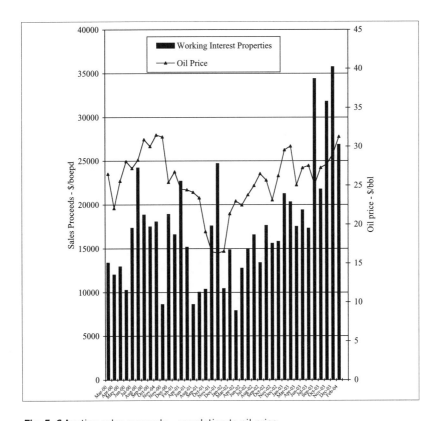

Fig. 5–6 Auction sales proceeds—correlation to oil price

Marketing Options

Brokers

Brokers are frequently chosen to market property for several reasons:

- Brokers are in the business of selling property, and are perceived by many companies to have processes and contacts that are better than in-house methods and information. This is true particularly when in-house sales are infrequent, resulting in no continuity between sales programs and personnel.

- Most brokers are able to cast a wide net when contacting potential buyers, because they maintain constant contact with the buyer community.

- When a buyer is selected at the conclusion of the bidding process, it is helpful to have a third party in the middle to run interference or *test the water* with proposals floated by either side if negotiating becomes tenuous.

- Some brokers have an affiliation with financial firms that can provide a third-party *fairness opinion*. This opinion must be rendered for certain publicly traded companies to validate that the sales price obtained for a rather large property sale fairly compensates the stockholders. When a fairness opinion is needed, it is convenient and cost effective to use a firm that can both market the properties and issue this opinion.

Companies may need to put certain limitations on a broker's activities. The seller will discuss these boundaries early in the process and create guidelines that will insure that the processes used by the broker are consistent with company policy. A few of the reasons for this follow:

- The seller may specify that the buyer must be of a minimum size as an indicator of financial strength for a protective measure. For example, the property may have a high abandonment cost with a lease that stipulates if the operator does not plug the field properly, the responsibility will revert back to the original leaseholder.
- The seller may review the broker's list of potential buyers and eliminate companies with whom they have had past differences, unpaid bills, or unresolved negotiations.
- The seller may want all bids sent to the company internally rather than having the bids sent to the broker, to maintain control over that critical portion of the process.
- The seller may ask that the broker abstain from participating in negotiations. The company may believe that the best results will be obtained when negotiations are handled in-house.

The marketing process is long and has numerous phases. There are many opportunities for the process to be interrupted, and, if just one potential buyer becomes sufficiently disgruntled, the sale of the properties can be involved in litigation that is costly in terms of time and dollars. Litigation or even the threat of litigation can totally derail the efforts of the seller to divest the property and taints everyone involved in the effort.

Brokers work diligently to deliver the desired results. Experienced brokers are motivated by the need to develop and maintain a reputation in the industry for being an ethical representative of the seller, as well as for being able to obtain property at a premium price for a qualified buyer.

In-House to Industry

There are a number of reasons that companies, particularly large ones, choose to market their fields themselves, including:

- Large firms have staffing levels that can absorb the need for ad hoc teams that will be temporarily assigned to sales programs on a periodic basis.
- Staff with the necessary expertise can be maintained within a company in between sales programs.
- The timeline of a sales effort can be accelerated, slowed down, or even terminated without breaking a contract with an auction company or broker if company objectives change.
- The company may want to maintain complete control of the information that is given to the buyers and the process as a whole.
- There is a perception that if commissions are not paid for an auction or to a broker, that the effort has no incremental cost to the company.

The trend at this time for some large companies, however, is to not market in-house for the following reasons:

- Staffing levels are bare-bones and the flexibility of the past is gone.
- The time and space consumed by a large sales effort is significant.
- The marketing environment is becoming more specialized with the use of electronic media.
- This activity is an important function with a high risk if it is not optimized.
- Divestment activity is rarely a priority for management.

Solicit Co-Owners

There are circumstances when a property owner believes that field co-owners may be the ideal buyers. In these cases, the owner will simply solicit an offer from the co-owners and negotiate toward an acceptable price and contract terms if they are receptive. Examples of when this may occur are:

- The field is in an isolated area.
- The field has unique litigation or abandonment problems.
- The co-owner operates in the area at low cost.
- The co-owner has been particularly aggressive in purchasing adjacent property.

Experience has shown, however, that even when it is a logical course of action to solicit an offer from co-owners first, it is not far-fetched to get the best price from a company that has no previous ownership or experience in the area. The rationale for this situation is that the co-owner knows the risks, potential costs, and downsides of the asset far better than other potential buyers. The buyer that is not a co-owner may overvalue the property because of a lack of awareness, thus being unable to incorporate the risks into the analysis.

THE DIVESTMENT PROCESS 6

Activity Timeline

Once the decision to sell an asset has been made, it can take six months to a year or more to complete the process. From start to finish, the necessary activities include, but are not limited to, the following:

- Review the characteristics of the asset.
- Determine the current tax basis and book value of the property.
- Perform the engineering analysis and economic evaluation.
- Quantify exploration drilling potential and upside performance reserves.
- Estimate the minimum acceptable sales price and financial impacts.
- Select the marketing approach.
- Identify the desired bidder qualifications.
- Obtain co-owner permission to show confidential data.

- Identify limitations on showing seismic data to potential buyers.
- Define the abandonment or environmental bonding requirements.
- Gather well bore schematics.
- Write an explanation of the operating plan of the property.
- Make land and lease summary reports with land plats.
- Tabluate gas imbalances by property.
- Note the marketing obligations and options for produced products.
- Consolidate the information for the data room.
- Package the information for online or CD-ROM presentation.
- Contact the potential buyers and make appointments.
- Set up the data room and seismic workstations.
- Manage the data room (or other data transfer options).
- Present the upside opportunities.
- Accept and review offers.
- Calculate the financial impacts.
- Select the best offer.
- Arrange for field trips and due diligence.
- Negotiate price and contract terms.
- Obtain management approvals.
- Contact co-owners with preferential rights to purchase.
- Obtain consents to assign from landowners.
- Close the transactions with buyers.
- Advise the unsuccessful bidders.
- Transfer all the files to buyer, both from the office and field locations.

- Contact regulatory agencies with a change-of-operator notice.
- Determine final recapitulation payment to buyer after closeout of accounting.

Depending on the marketing approach that is used, the above steps will vary somewhat. Participants in the process will engage at different times and may engage more than once. The above is only partially a sequential process, and it is common for overlap in activity to occur where possible so the timeline can be compressed.

Data Room Requirements

Before 1995, a data room was the only method used to convey information to potential buyers. It would be stocked with hardcopy reports containing the information needed to assess the offered property's risks, upside opportunities, and current value. The reports included ownership, land and legal, reservoir, production, geological, well, field, accounting, and marketing data. At times a subset of the critical information would be precopied for distribution. The potential buyers would be required to copy any additional data they wanted, trace structure or formation thickness maps of interest, and in general cart large volumes of paper from the data room at the end of the day.

The data room has two components. The first is the basic information described previously that was needed to determine field value and the associated risks. The second is a presentation that gives the seller the opportunity to show how field value could be increased, both operationally and by drilling. At times an exploration play is involved if enough leasehold acreage is present or a deeper play in the area is untested. This upside component is critical, as the buyer who *sees the vision* and values the potential more than the competition will likely be the high bidder and acquire the field. The upside value assessment carries the higher risk as it is more speculative than the first component.

In recent years, the availability of 3D seismic data has added an additional level of complexity to the data room process with the myriad of licensing terms associated with the various vintages of 3D data. Now, sellers must determine:

- What 3D data the potential buyer can see (both hardcopy and on the workstation)
- What interpretations the buyer can trace (whether off the screen or off prints)
- To what extent the buyer can *drive* the workstation to view the seismic
- What hardcopy material (maps, cross-sections, or lines) a buyer can take from the data room

Although showing 3D data has complicated matters somewhat, it has opened up a world of new opportunity and value to the marketplace. Some reserves' accumulations that were once considered high risk now have enough technical support to be valued by a purchaser. Also, exploitation and exploration opportunities that are supported by 3D seismic data can be shown with a higher measure of certainty, and as a result the marketplace is paying more than it ever has for higher risk potential.

As the marketing choices become more diverse in the online era, the delivery of information to the buyer universe has become more efficient. CDs containing huge amounts of information have replaced or drastically changed the paper data room. Broker or owner Web sites are also an alternative to visiting data rooms. The challenge to some buyers at this time appears to be finding an easy way to assimilate and print the reams of paper and large scanned maps that are available in the electronic media.

The characteristics of a data room have evolved significantly in the past decade from an approach that was relatively passive to one in which far more activity is undertaken (see Fig. 6–1).

	1990	2000
Location of data rooms	Company in-house	Broker
Data medium	Hardcopy	CDs, Web sites
Technology	Low tech	High tech
Approach	Seller oriented	Buyer friendly
Buyer travels with	Portable copy machine	Geophysical data

Fig. 6–1 Evolution of seller data

Although the alternatives for presenting information in the marketplace are more numerous than ever before, there will always be a need for personal contact between sellers and buyers to discuss many matters, including upside potential, terms and conditions, and ongoing litigation.

Third-Party Reserves Report

The need for a consultant's engineering report can be debated. At times, both the buyer and seller need one, and at other times, both parties may be hampered by having one. Often one party wants a report and the other does not.

Some of the reasons an engineering report is helpful are:

- It is a roadmap that can quickly lead the buyer to the principle areas of value and show the relative uncertainty of the reserve accumulations.

- A third-party view gives the buyers a benchmark for comparison of in-house values.

- It helps the seller to understand what the market perception will be for the property being offered. Frequently this is a lower value than the seller hopes to promote.
- It speeds the timeline to closing as buyers who need financing must have a third-party report to serve as the basis for loan value.
- If a property that was marketed as a portion of a package cannot be included at closing (perhaps because of a lease title failure), the report functions as an impartial determinant of value when the field is removed from the transaction.
- The firm preparing the report typically has a staff that is at the forefront of the technology being used for properties in the area.

Conversely, an engineering report may not be provided because:

- It can be costly and the time needed to engage the consultant and perform the work might delay the front end of the marketing timeline.
- The report may limit the value placed on the property by the buyer.
- The report may downgrade or eliminate reserves or upside potential that the seller wants the market to value.
- The report may limit the amount the buyer can borrow against the property, even if the buyer takes a more robust view of the reserves.

One hurdle to hiring a consultant to provide a third-party opinion can be cost. Many companies have gotten beyond this by staging the review and having defined decision points that identify when the analysis will be terminated, continued by the consultant, or taken over by in-house staff. As this approach has built in periodic review points, it works well to keep management informed of the progress of the ongoing analysis.

The factors to consider when deciding to obtain a third-party engineering report include the seller's timeline, the seller's viewpoint of property value relative to the data, the scope of the property in the package, and the cost.

Another perspective on this decision is that the presence of a report will lessen buyer risk, which leads the buyer to pay more for the assets. One risk to the seller that is mentioned at times is that the report may lessen the chance that a buyer will overpay for the assets. This is not always a valid concern, as typically the assumptions and calculations are made in a consistent manner based on SEC guidelines. The seller must balance these factors and make the decision on which route will optimize the sales effort.

Generally, the engineering report focuses on proved reserves and the upside that can be derived from those reserves. Additional documentation outside of the report is usually prepared to show the exploitation or exploration upside potential, so the basis for upside value is presented to the marketplace in its entirety by the seller. The degree to which the buyers accept this upside is frequently the factor that separates the successful bidder from the other players.

THE ACQUISITION PROCESS 7

Steps to the Process

The approach to an acquisition analysis varies by company and is greatly influenced by the type of asset to be evaluated. However, as discussed in this section, there are common elements that apply to nearly all acquisition attempts.

Step 1—Perform a quick screen:

- What is the attraction to buy the asset?
  - Growth opportunities
  - Cost saving measures
  - Synergies with neighboring operations
- What is the situation?
  - Data room setup; deadlines and personnel requirements
  - Willingness of seller to sell
  - Co-owner activity, preferential rights
  - Potential competition

Step 2—Decide to move forward:

- Review land and lease
- Define most realistic depletion scenario
- Determine well completions (tubing sizes, back pressure, compression, sand control)
- Determine reserves (all categories)
- Schedule the production until depletion
- Forecast operating costs and investments
- Validate the financials

Step 3—Capture the exploration potential:

- Understand the growth scenario

Step 4—Analyze sensitivities to the base case:

- Unrisked and risked reserves volumes
- Pricing cases, hedging
- Investments (cost and timing)
- Well completion options
- Operating costs (fixed and variable)
- Abandonment and environmental costs

Step 5—Company impacts:

- Effect on existing operations and overhead
- Competitiveness with other company investments

Step 6—Perform bid analysis:

- Check economic parameters for various bids
- Determine the discount rate
- Understand tax impact and write-offs
- Project UOP rate and annual earnings

Step 7—Review other impact items:

- Revenues and costs
- Existing gas imbalances
- Off-lease production processing
- Uplift for marketing operations
- Operations, including whether property is operated by the company or by others
- Well integrity and mechanical risk
- Production
- Past three years performance compared to projections
- Economic limit projection

Proactive and Reactive Approaches

Although there are many ways by which a property may become a target for acquisition, the process can be separated into two situations:

1. The acquiring company does its homework privately to proactively identify a property as a target while the owner of the property is unaware of the effort.
2. The seller has chosen to sell a property based on internal criteria and makes it available to some segment of the industry that reacts to the offering.

Each of the approaches is used regularly and each results in successful transactions. Sometimes a sale starts with the first process and is ultimately blended into the second when a seller wants to be sure market value is being paid. Most successful buyers have built profitable portfolios by merging both approaches into their business plans.

The proactive approach

The proactive approach is used when a buyer makes an unsolicited offer to the seller, followed by negotiation of price and terms. The mode is used by companies who have the resources to do the preoffer analysis of new properties with no guarantee that a seller will want to sell, much less that a transaction will result.

Companies that make acquisitions in this manner are usually very competent exploiters of property and pick their targets based upon drilling opportunities or synergies in operations. There is generally no industry competition in this approach, which is also characterized by strong relationships between buyer and seller leading to repeat business. The challenge is being able to match the seller's minimum price and still make an acceptable profit.

This method also includes attempts to buy out co-owners. Many situations lead a company to exercising this course of action. For example:

- If a property is burdened by higher-than-usual royalty interests, the economic limit will also be higher than usual. These royalties can be purchased to lower the economic limit of the working interest volumes. This in turn would result in the conversion of previously unrecoverable volumes into proved reserves, which would add a second level of profitability to the transaction.

- When co-owners buy out a partner's working interests, it spreads the fixed technical overhead over more volume, thus increasing profitability and impact.

- The purchase of additional equity in property that is currently owned is recognized to be a comparatively low-risk endeavor, because one assumes a thorough knowledge of the operations and reserve base by the buyer.

The reactive approach

The reactive approach to property selection is more common in the industry. The company decides which assets to sell and may hire assistance for the marketing effort. Depending on the approach, the companies themselves, brokers, auctioneers, or online agents will send up to hundreds of notices to the industry advising of a property that is for sale on a regular basis. Because companies spend less time in data rooms than in the past as a result of the new marketing tools that are now available, the number of companies that are contacted has risen substantially.

At times this process is very selective, however, as the seller may contact a limited number of companies for efficiency and speed. Much of this early effort is devoted to communicating with the potential buyer pool by phone to assess buyer interest.

Data Room Visit

The team that is assembled for the data room should be an energetic group of professionals that includes all necessary skill sets for a thorough understanding of the property. They should know exactly what information they are expected to acquire and what will be done with it by whom when the team returns to the office after the visit. While in the data room, the emphasis of all individuals should be on:

- Locating and capturing the information that they have responsibility to gather
- Identifying the risks or opportunities associated with the subject matter that defines why they are on the team

The data room trip is not the time or place to perform an evaluation analysis, even if time permits. If a team member has exhausted the search for information and has asked the seller all the appropriate questions, he or she should pitch in to assist other team members who may be overwhelmed with their tasks. After the trip is concluded, the team meets again back at the office to discuss the characteristics of the property, the risks that each team member may have noted, staff that are needed that were not on the initial team, the sequence of the evaluation, additional resources such as technical support, and the timeline with back-away decision points.

Reserves Assessments

Although every phase of the acquisition process is important, the determination of recoverable reserves by the buyer is arguably the most important of all. If every participant in the process does outstanding work but the reserves are overstated, it is usually impossible to salvage profitability from the transaction.

The easy solution to this is to be conservative in calculating the reserves, but this approach will likely result in not making any acquisitions. Success belongs to those who can best identify the total resource base of an asset, and then determine how to maximize recovery from that resource base by the most efficient and timely use of technology, investment capital, optimization of operations, and project management.

Years ago, an aggressive bid valuing proved reserves was enough to be successful in the acquisitions arena. However, the competitiveness of making acquisitions has intensified in recent years because of:

- The improvements in technology in each of the oil-field disciplines
- Focused and highly skilled company acquisition teams
- Fewer quality fields being offered in the marketplace
- The general availability of borrowed capital

As a result, the valuing of probable and even possible reserves is more common and, in fact, is essential for success. A buyer needs to *see the vision* of total field depletion, not just the proved reserves with easily quantified upside. This pushes the risk envelope of the buyers out farther and farther, but the skill sets and tools brought to bear to the evaluations phase of the analysis far exceeds the capabilities that were available in the past.

The need to evaluate the complete asset has been a driver for the use of consultants to assist the project team. At times the acquisitions team is not expert in the geographic area of the offering. To bridge the information gap, add resources to the effort, and provide an unbiased, *noncompany* viewpoint to the analysis, it is helpful to hire consultants. Many consultants have frequently worked on acquisitions in the producing area of interest over the years. Their integrated teams are composed of engineers, geoscientists, and geophysicists, and they bring a multidisciplinary experience to the process that is very helpful, if not indispensable.

Financing Options

Acquisitions are either internally funded or financed by a lender. When a company has the necessary funds to make an acquisition, it places that company in an advantageous position against those who must borrow the money. Sellers prefer to sell, given the choice, to a company that has the money because:

- Their timeline to closing is not impacted by the lending institution.
- The transaction is not at risk if the buyer cannot obtain financing.
- A third-party reserves report is generally not needed (this could extend the timeline to closing if a report is not requested until a deal is negotiated).

Funds for acquisitions are available from a number of sources. Banks may be the most conservative of them all, and will loan half of the proved reserves value as a rule of thumb. In exchange for only being able to borrow a portion of the value of the acquisition, the interest rate offered by a bank may be the lowest of all the options that are available for financing.

The most costly source of funds may be mezzanine financing, in which the lender will take part of the deal in the form of equity, with perhaps a back-in after payout or some multiple of payout or royalty. Mezzanine financing is very common and is used extensively by smaller companies that may not have the collateral or market value to obtain bank loans of the necessary size.

Product Price Considerations

Although the product price forecast that is used in the evaluation is obviously important, the historical purchase price paid per barrel equivalent follows a trend that mirrors but does not match the realized wellhead price cycle.

Oil and gas price plots can be obtained from a number of sources. A review of the product price trends compared to the prices paid for divested hydrocarbons shows the general correlation that the peaks of the transaction price plot will lag behind the peaks of the price plot by several months. This would be expected, as buyers will not increase their evaluation price decks on the strength of a brief ramp-up in wellhead prices.

However, it is also apparent that when product price falls, transaction prices fall rather quickly, even though most transactions have lives that are long enough to mitigate short-term price drops. The reason for the quick reaction to a drop in product price is that if the near-term profitability of a transaction falls short of expectations, it is very difficult to attain the preacquisition profit expectation given the discounting of cash flow in the economic analysis.

Another observation from the pricing comparison is that, although the wellhead price variation in any two-year period ranges from $0 to as much as $20 per barrel, the price paid per barrel variation was at most $2 per barrel in the same period. From 1982 to 2002, the price for all deals remained within a band between $4.30 per barrel and $8.60 per barrel, while the wellhead price remained within a much wider band of $10 to $40 per barrel.

The data are similar for transactions that are primarily gas. Although high gas wellhead price swings have occurred (most notably in 2001 when prices passed $9 per one thousand cubic feet [Mcf]), the average price paid of $1.25 per one thousand cubic feet equivalent (Mcfe) that year is very conservative in comparison. At the low end of the price history in 1992, when the average wellhead price was just $1.00 per Mcf, sales of gas reserves still managed to average $0.70 per Mcfe.

The previous discussion on market multiples for each product excludes transactions for distressed property, in which the reserves sold for minimal value because of near-term end-of-life abandonment costs. If the abandonment costs were added to the purchase cost and used as a proxy for price paid, the multiples would be expected to fall in the reported band of values.

Acknowledging the relative volatility of the comparative price information, it is obvious that the price deck used in the evaluation to sell or buy a property takes on far greater importance than just being one of many parameters that are used to establish market value. The company that makes the best guesses at wellhead prices will make astute portfolio management decisions and be more competitive than the other active producers in the acquisition and divestiture (A&D) marketplace.

Modest wellhead price changes can have a significant impact on profitability. Assume in the base case that gas sells for $2.50/Mcf in the current year. If the lifting cost is $0.50/Mcf and the combined severance taxes and royalty is 25%, then the before-tax profit is $1.37/Mcf. If the price rises by 50% to $3.75/Mcf, the profit increases nearly 70% to $2.31/Mcf. Similarly, on the downside, if the price falls by 50% to $1.25/Mcf, the profit decreases nearly 70% to $0.44/Mcf.

This illustration shows the impact of price volatility. For example, the market for heavy crude asset sales is limited and shows low market multiples. A drop in wellhead price by a few dollars can quickly eliminate any profit on a heavy oil acquisition. In contrast, an unexpected increase in price will create additional profit that would be difficult to achieve otherwise. Any prudent acquisition will reach payout in an accelerated time frame when prices rise unexpectedly for a sustained period of time.

Major cycles in price occur in the industry every few years, and sometimes more often. Most properties experience several of them during their life cycle. The revenue swing that a property experiences in relation to the predicted acquisition analysis can be huge and unsettling (particularly during the low periods). This is why cost control is so important. A high cost structure can easily lead to an evaporation of profits during low price periods.

UNIQUE LIFE CYCLE RISKS   8

Each producing property has a life cycle that is unique in its length, risk, cost, recovery, and value. However different these characteristics may be, each property passes through the same stages from discovery to depletion. The time period may be as short as 5 years or as long as 100 years. The stages and their impact on the acquisition are discussed in this chapter.

Exploration

There is great uncertainty in the value of a discovery prior to delineation. If the prospect is encouraging, it is rare that a participant will divest at a price that reflects the relatively low proved reserve volume, even when the myriad of risks that are associated with this stage of the life cycle are recognized. The risked reserve value will be a fraction of the total resource potential, thus sales of discoveries are uncommon unless the development plan does not fit the owner's company strategy.

Delineation

After a discovery is delineated, a sale is more likely to occur. At this stage, the capital costs are framed and the operating scenario is generated, thus the cost side of the analysis is reasonably defined. The reserve picture and expected production rate are clearer than at the time of discovery, but still carry great uncertainty. It is at this stage that the co-owners decide individually if the anticipated reserves and cost will combine to create a property that is a good addition to the company portfolio. If not, the property may be sold in lieu of spending the capital that is necessary to produce the discovery.

Development

In this period, while the projected long-term producing rates and drive mechanisms are better understood, the range of the ultimate reserve volume is also better defined. A working interest in properties at this stage is very valuable and marketable, because:

- The initial *flush production* is usually at a high rate.
- The risk of cost overruns is limited.
- The reserves are known well enough that a terrible overestimation is unlikely.
- The abandonment obligation is far in the future and discounted heavily or ignored.
- The property probably has rework potential and some drilling to provide upside value.
- There may even be exploration potential on the flanks or deeper on structure.

The few properties that are sold at this stage of the life cycle get high multiples and generally reward the seller with a good return on sunk costs.

Production

After a field has produced for a number of years and the operating cost structure, drive mechanism, seismic data, and ultimate recovery are understood, sales become more common. These properties would generate high interest if marketed, as they are usually very profitable and have upside potential. Fields are generally not sold at this stage of their life cycle, however, since their attributes place them in the core area of most company portfolios. Thus, proactive buyers are the more common purchasers of property in this mid-life stage.

As fields pass into the later stages of profitability, most are characterized by several of the following attributes:

- The production is much lower than in earlier years and is from fewer wells.
- The number of inactive wells in the field outnumber the producing wells.
- The field value is moderate relative to abandonment cost.
- Any environmental problems that have been created are apparent.
- Only limited or risky drilling upside remains (the low hanging fruit has been harvested).
- The fixed operating costs per barrel have risen to the extent that the field may not be a desired part of a company's portfolio.

These attributes characterize the majority of fields that are sold in a reactive manner. They have positive value and contain enough upside potential to attract a suite of aggressive operators and companies with the better exploitation resources.

If a company has a number of mature fields in an operating area with little chance for quality improvement, it should trigger a review by the company to determine if an exit strategy for the whole area is best while decent value can be obtained for the property.

Redevelopment

Most fields experience revitalization after the initially drilled wells have reached the advanced stages of depletion. This well work is normally undertaken before the field's financial metrics become so depressed that the asset is considered to be a divestiture candidate. The two most common reasons for revitalization are:

1. New seismic data is obtained that has better quality than the seismic data that was available at discovery. An undrilled reservoir area is noted and subsequently exploited.

2. Reservoir production and performance data are analyzed and indicate the need for additional wells to increase or accelerate recovery.

Revitalization efforts are frequently very profitable as the infrastructure to support the new production is in place. Many companies who are proactive acquirers look for fields that have not been revitalized, because fields in this stage of their life cycle have consistently been fertile ground for finding quality locations. The downside of these locations is that the reserves target per well is not normally as large as the original development, simply because the largest reservoirs have been produced by this time.

Sunset Period

The fields in the *sunset* stage of their life cycle are nearing abandonment and have poor chances of regaining profitability. The fields may have just one or two producers and carry a huge risk if one of the key remaining wells goes off production, as it is usually not economical to rework. The fields are naturally difficult to sell individually because of their minimal value. As a result, these fields are generally marketed and sold as a part of a larger package that contains better fields to attract a buyer and support the transaction.

VALUATION METHODOLOGIES 9

Fair Market Value

There are a number of definitions in the industry for fair market value. One that reflects the actual price paid in property acquisitions and divestitures is as follows:

Fair market value is the price a willing buyer will pay a willing seller, both having full knowledge of the facts, neither being under pressure to make the deal, with the property being available to a representative portion of the market for a reasonable period of time.

Another, more academic definition that excludes the "investment value" a company may incorporate into the first definition is:

Fair market value is the average value calculated by a mix of competent engineers and geoscientists who have no interest in the property, its sale or its purchase.

This second definition excludes the potential synergies, applications of unique technology, or any other factors that would make the property more valuable to one buyer versus another. The premium paid by the successful buyer represents the "investment value" that it assigns to the property. The successful bidder is generally the company with the highest investment value. Because the investment value is greater than the fair market value, one does not know the fair market value of a property (based on the second definition), nor the premium paid, based on sales prices reported by the media.

The determination of estimated fair market value in anticipation of a competitive sale is difficult and is frequently lower than the price that is offered by prospective buyers. The market is dynamic, and each buyer has its own motivation and bias. It is important to note that market value is an estimated number prior to the signing of an actual purchase and sale agreement.

The Society of Petroleum Evaluation Engineers has recently published a monograph entitled *Perspectives on the Fair Market Value of Oil and Gas Interests* that addresses various aspects of the calculation of fair market value. The monograph was prepared with input from a large cross-section of experienced evaluation and reserves engineers who are well known in the industry. It gives guidance on evaluation procedures, the nature of risk and uncertainty, adjustments to manage these factors in an evaluation, and the quantification of fair market value influenced by market information when available. It is a very useful resource that can be consulted for detailed information and example problems on this topic.

The concept of strategic value is discussed early in the monograph. An understanding of strategic value is crucial to success when purchasing producing property on a competitive basis. Whereas *fair market value* is generally calculated by evaluators using industry standard procedures, *strategic value* is the premium that a buyer is willing to pay for the unique reason that the property in play is a particularly good fit to its organization. The company that is astute enough and aggressive enough to identify and pay for strategic value will experience success more often than the company that does not.

Seller Versus Buyer Viewpoints

There are a number of reasons why a buyer can have a different, more optimistic viewpoint of a property's value as compared to the seller. For a mismatch of perceived value to lead to a transaction, at least one of these factors needs to occur. They are:

- The buyer may have a lower economic hurdle rate for investment opportunities; thus the buyer will include more reserves in the acquisition analysis.
- The buyer may use lower cost estimates for future investments based on riskier, streamlined procedures for the same well work.
- The seller may have exploration and exploitation potential valued at the revenue that could be derived from farming out the opportunity, whereas the buyer may drill the prospects.
- The buyer may be confident in the application of new technology.
- The buyer may have a more optimistic view of the reserve volume or field life because of a lack of negative information on the property.
- The buyer frequently is a smaller company having a lower cost structure (both in the field and the office) and as a result can squeeze a higher profit margin from production volumes.
- The seller may see more risk in the continued operations scenario.
- The forecast of product pricing is always different between the parties, and will create a value difference if other aspects of the field evaluation and operations are the same.

- The buyer may place a high strategic value on the asset.
- The buyer may have offset operations that will lower the operating costs in the area.

There are many other reasons why the fair market value of a buyer and seller can differ enough to support a transaction, but these are the principal ones.

Market Multiples – Asset Acquisitions

The sales price per barrel equivalent of reserves, the sales price per barrel equivalent produced per day, or the sales price expressed as a multiple of the most recent year of net cash flow are three commonly used multiples to estimate the fair market value of a producing property. For example, the following ratios were typical of deals reported during the late 1990s involving fields located along the Gulf Coast or in the Gulf of Mexico that were in the mid-life period of their life cycle with moderate upside potential during a period of *average* wellhead price:

- $5–$7 per boe of reserves
- $12,000–$18,000 per boepd of production
- 3 to 5 years of cash flow

As wellhead gas prices soared in a year spanning parts of 2000 and 2001, these multiples increased into previously uncharted territory. The profit margin per unit of production doubled and, in some cases, tripled, rendering these historical multiples useless as predictive benchmarks. The influence of hedging and optimistic expectations on price resulted in a situation where predicting what price a marketed field would sell for was a guess at best.

After the run-up in price was judged to be relatively permanent and not an anomaly, price multiples did stabilize higher as shown in Table 9–1 and Figures 9–1A and 9–1B. The average price paid per boe increased from a 20-year historical average of $4.52/boe to $5.89/boe from 2000 to 2004. The average price paid per boe produced per day increased from $10,439/boepd to $24,876/boepd. The price paid for production jumped a much higher percentage than the price paid for reserves. Reasons for this may include:

- Application of new completion technology has increased production rates.
- Accelerated rates lower the reserves risk and shorten the time of deferment.
- Advanced 3D seismic data and interpretation allows for faster implementation of drilling programs and exploitation of smaller reserve targets.

Table 9–1 Relative cost of cash-based transactions by time period

	1979 - 1999	2000 - 2004
Number of deals	1,402	199
Total acquisition cost - $MM	58,718	17,869
Total reserves - MMboe	12,998	3,035
Average size deal - Mboe	9,271	15,251
Average cost - $/boe	4.52	5.89
Production data available:		
Number of deals	749	204
Total acquisition cost - $MM	39,095	19,901
Total production - Mboepd	3,745	800
Average size deal - boepd	5,000	3,922
Average cost - $/boepd	10,439	24,876

Generally, reserves data is proved reserves only.
Boe = barrels of oil equivalent.
boepd = barrels of oil equivalent per day; M = thousand; MM = million.

Valuation Methodologies

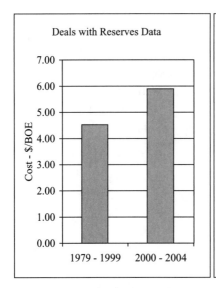 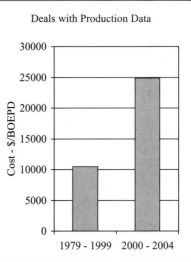

Fig. 9–1A and *9–1B* Comparison of price paid by time period

The market multiples from the late 1990s were taken from published sources. It must be noted that usually the large transactions are made public with the data that is needed to derive these multiples. The smaller transactions, of which there are many, do not get reported as frequently because they are not as noteworthy and rarely have all the data reported that is needed to make the calculations.

The multiples can vary greatly from the historic range when a single sale is reviewed, as the hypothetical illustration in Table 9–2 shows. Many factors can cause the remaining reserves, current production, and projected cash flow to create multiples that are not consistent with one another.

A seller of low-caliber assets should be careful and not assume that market multiples are always representative of value. Many low-quality fields may have positive cash flow, but the present value of the property is negligible because of end-of-life costs. These fields have such a low value that they may be assigned to the buyer for no consideration. Deals of this nature rarely make the headlines and have multiples of zero because the selling price is $0 in assignments.

Table 9–2 Market multiple calculations

Field Characteristics	Multiple	Market Value
Production—500 boepd	$12,000 to $18,000/boepd	$6.0 million to $9.0 million
Reserves — 1 million boe (5+ years rate life)	$5.00 to $7.00/boe	$5.0 million to $7.0 million
Cash flow—$1.5 million/year ($8/boe profit)	3 to 5 years	$4.5 million to $7.5 million

boe = barrels of oil equivalent; boepd = barrels of oil equivalent per day.

When the market value of an asset is estimated, a range of figures is calculated rather than a single value figure. In this example, the market value of this property would be between $5.5 million and $7.5 million.

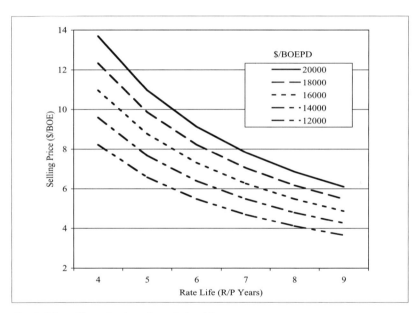

Fig. 9–2 Rate life and sales price relationships

Figure 9–2 shows the relationship between the reserves price multiple ($/boe) and the rate life (reserves volume divided by the annual production, commonly referred to as the R/P). The lines on the plot represent varying prices paid expressed as the parameter $/boepd. These show how the rate of recovery impacts the sales price.

Valuation Methodologies 77

The faster the recovery (the lower the R/P) of a given reservoir, the higher the price (i.e., the cost of deferment and recovery uncertainty is reduced). The plot shows that for a constant R/P, as reserves quality decreases the deal metrics also decrease. Similarly, for a constant reserves price multiple, as reservoir longevity and rate life increases, the production price multiple also increases.

Figure 9–3 shows how the market multiple parameters change from the Permian Basin, with its low decline rates, to the Gulf of Mexico, with its high decline rates.

A. Permian Basin:

    Field parameters: 5 MMbbl reserves, R/P = 10, annual production = 0.5 MMbbl.

    Asset is sold for $5/bbl, or $25 million. The production multiple is $18,250/boepd.

    The high rate life of 10 is associated with a high deferment cost. Thus the reserves multiple ($5/bbl) trends low and the production multiple ($18,250) trends high.

B. Gulf of Mexico:

    Field parameters: 5 MMbbl reserves, R/P = 5, annual production = 1.0 MMbbl.

    Asset is sold for $7/bbl, or $35 million. The production multiple is $12,780/boepd.

    The high rate life of 5 is associated with a low deferment cost. Thus the reserves multiple ($7/bbl) trends high and the production multiple ($12,780) trends low.

bbl = barrels; boepd = barrels of oil equivalent per day; MMbbl = million barrels; R/P = reserves volume divided by the annual production.

Which is the better property to purchase for your portfolio?

This illustration shows that one property having a certain reserves multiple at a given production rate will have a lower production multiple when compared to another property that has the same reserves volume but is producing at a higher rate. It is important to understand that when extrapolating transaction parameters from one sale to another it is crucial to gather as much information as possible to understand the assets.

Fig. 9–3 Comparison of market multiples for different locations

In closing, market multiples are used often for predictive purposes. Great caution should be used when making these predictions—the public data must be massaged to account for the differences in the property that is being evaluated versus the property characteristics that were reported. Frequently, nonproducing assets such as exploration leases, mid-stream or downstream businesses, or favorable long-term contracts will elevate a deal multiple, making it a benchmark that would be impossible to match by an asset transaction that did not have similar extenuating circumstances.

Market Multiples – Historical Figures

In general, the larger a transaction is the higher the price paid per barrel equivalent. Larger deals tend to have better quality fields, more reserves, a higher likelihood of reserves growth over time, and exploitation drilling potential. The sale statistics by transaction size shown in Table 9–3 include transactions from less than $10 million to transactions up to $100 billion. The table includes 3,800 deals transacted since 1980 for which sufficient data were reported.

The mega-mergers head up the list with deals more than $10 billion in size and averaging $19.88/boe! All transactions from $100 billion to $100 million in the first three size categories amounted to just 12% of the total deal count, but they accounted for 93% of the total cost and 86% of the total reserves that changed hands. As the size of the deal fell to less than $10 million, the price paid dropped to a low of $4.89/boe. Given this dramatic range of prices paid per barrel, it is obvious that the concept of strategic value noted earlier greatly influences merger acquisition prices.

Each of the deals noted in Table 9-3 are shown in the cumulative transaction count plot in Figure 9–4. It shows a smooth distribution of deal sizes with approximately 75% of them falling within a size range of $1 million to $100 million in value.

Table 9-3 Reported sale statistics by transaction size

Deal type	Price range $B	Distribution Count	%	Distribution $B	%
Mega mergers	100 - 10	9	0%	305	46%
Large mergers	10 - 1	60	2%	206	31%
Mega deals	1 - 0.1	399	11%	110	17%
Big deals	0.1 - .01	1275	34%	40	6%
Small deals	.01 - 0	2056	54%	6.8	1%
		3799		667.8	

Deal type	Price range $B	Reserves Analysis BBOE	% BOE	$/BOE
Mega mergers	100 - 10	15.3	31%	19.88
Large mergers	10 - 1	16.6	34%	12.43
Mega deals	1 - 0.1	11.4	23%	9.65
Big deals	0.1 - .01	4.4	9%	9.08
Small deals	.01 - 0	1.4	3%	4.89
		49.1		13.6

Transactions from 1980 to June 2003

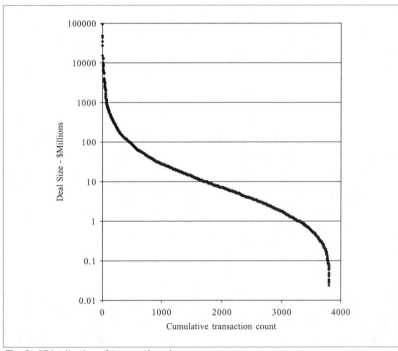

Fig. 9-4 Distribution of transaction size

Operational Concerns

Although the numbers that define reserves and profitability are a decisive part of the evaluation for a buyer, the characteristics of field operations must also play a role in the determination of bid price and the decision to walk away from the opportunity or to pursue it. The complexity of field operations must be understood and communicated to management.

The quality of existing operations is important—the current operator expertise and reputation, capital efficiency record, concern for environmental regulations, apparent hurdle rate for investments, and efforts to control costs needs to be understood. It may be unreasonable to assume that it is possible to operate more efficiently than the *best in class* in a certain geographic area. Conversely, to buy from an operator who is very inefficient may provide a gold mine of opportunity.

DETERMINING THE PRICE   10

Impact of Various Factors

There are many tangible and intangible factors that impact the sales price. Some are related to the property, others relate to the circumstances of the buyer, some are a function of financial aspects of the deal, and others relate to the marketing effort.

Property characteristics that have an impact on getting a high sales price, in general order of importance, are:

- Net revenue interest to working interest ratio (want high)
- Reserve volume and reserves to production ratio (want high)
- Proved reserve volume relative to potential volume (want high)
- Investments required to maximize recovery and the associated risk (want low)
- Product price trend (want rising or stable)
- Production rate (want stable or on moderate decline)

83

- Production volume risk (want evenly distributed between many wells)
- Reservoir volume risk (want evenly distributed between many zones)
- Upside potential and timing (want high and exploitable in near term)
- Lifting costs (want low)
- Producing life (want long)
- Condition of production facility (want in good shape)
- Deep drilling rights (want included in transaction)
- Abandonment cost and timing (want low and later)
- Product marketing (want unencumbered)
- Environmental and legal exposure (want low)

The discount rate used by the buyer in the analysis has a large impact on the price that is offered. The rate is inversely proportional to the price that can be paid (the higher the discount rate used, the lower the price paid). Although this relationship is simple to understand, it is not simple to use it to gauge the aggressiveness of any buyer relative to the competition. The reason is that some companies use a higher discount rate to impart risk into their cash flow, while other companies risk the reserve volumes and use a lower discount rate. Absent a thorough review of the reserve determination and economic analysis, it is difficult to make broad assumptions regarding the discount rates used by any company in an acquisition analysis.

The circumstances of the buyer have a significant impact on the aggressiveness that can be displayed. If a new core area or a competitive edge in an existing operating area can be gained with an acquisition, the buyer will be willing to pay more for a property than in other situations. If a company is looking for positive press clippings and analyst coverage in anticipation of an IPO, for example, it may push harder to acquire an asset with obvious synergy.

When a buyer can acquire operations and be in charge of the well work (timing, procedures, rig selection, and cost), that carries a value premium above the acquisition of the same interest if operations are not obtained. A nonoperator is always at an information and control disadvantage, causing nonoperated interests to be generally less attractive than operated interests. However, some companies thrive as exclusively nonoperated field participants because overhead can be kept low. These companies do a good job of evaluating their operators prior to making an acquisition and place a high degree of trust in them.

Funding sources vary greatly between companies. Those who rely on mezzanine financing pay high rates of interest. This reduces profitability and must be factored into the analysis. Large companies enjoy the more favorable and less costly circumstance where internally generated funds are available for acquisitions.

One other attribute of higher quality acquisitions is that the investment opportunities can be funded by the cash flow generated by the acquired property. When outside cash is not needed to fund the drilling and well work projects that will boost the production rate above preacquisition levels, the property is very attractive.

Multiple Field Packages

When a number of fields are combined and sold as a single package, the factors that need to be considered expand beyond those noted in the previous section.

For example, the mix of value in the package by field should be investigated to determine if all the value is in one key field and the rest of the fields are burdensome sunset properties. If so, the attractiveness of the core property must offset the condition of the distressed fields, as it may be impossible to spin off the poorer fields individually if the buyer does not wish to retain them.

On the opposite side of the spectrum, a package may have several fields that are high-quality properties. A package of this caliber will attract a large pool of buyers and sell for a high price multiple. Packages of this nature are rarely offered.

Another aspect of bidding on a package of properties is that the workload to evaluate the fields properly increases by some multiple over that of evaluating a single property. Additional resources are needed to manage this situation, which is made even more difficult when the seller's timeline is not adequately lengthened to accommodate the size of the package. In these cases, the evaluation team needs to be sized and managed from the beginning to insure a complete analysis of at least the primary fields in the package. Fields with low value may not get a detailed analysis, but should have a thorough review of their risks and end-of-life costs.

Sensitivity Analysis

Although a number of factors are merged to perform the economic analysis, some have a much higher impact and need to be worked to a greater degree of accuracy than others. Companies place high emphasis on proved reserves and production rate estimates, as well as the potential in the probable and possible categories. Cost estimation is also important, but is not worked as rigorously because it does not have as much impact in the evaluation results.

The following is an example of an analysis of the relative impact of changing parameters to a simple field evaluation.

The factors involved are:

- A field is producing 600 boepd with a 20% annual decline and 1 million boe reserves.
- A $750,000 well expenditure is scheduled in year one.
- The property costs $50,000 per year to operate.
- The present value of the field is $5.3 million.

Figure 10–1 shows the impact of a favorable 20% change to each of the listed variables individually. It shows that the reserves and production volumes are the most sensitive parameters, followed by the investment and operating costs.

In general, the relative importance of the factors shown in Figure 10–1 will not change if the property being evaluated is beyond the exploration and delineation phases of the life cycle, regardless of the asset that is evaluated. The company that can visualize the most certain path to the highest reserve volume and/or producing rates has a good chance of success. If that company also uses a competitive discount rate and price deck, it will have a very good

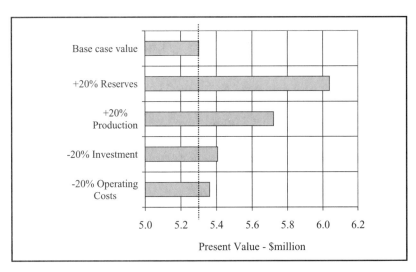

Fig. 10–1 Imact of primary evaluation variables

Determining the Price

likelihood of making the acquisition. If a company is not competitive on the reserve side, but can claim operating synergies because of existing production in the area, it is unlikely that the savings will create enough revenue to beat the company that identifies the upside reserve scenario.

The 20% variance was chosen as an arbitrary figure, recognizing that it is not outside the bounds of reason to have the ultimate reserve, production rate, or cost vary by at least that much from the input after the acquisition is closed. In practice, the variance that is chosen for sensitivity analysis should be logically matched to the actual case characteristics (rather than using an arbitrary figure such as 20%), so the analysis is as relevant as possible.

Product Price

Price forecasts are generally set by company policy independent of the impact they may have on the analysis. Regardless of the corporate view of price over the long term, the company that maintains the best historical perspective toward price cycles and aggressively pursues acquisitions while prices are low will have an advantage in the long run.

The factor that levels the playing field for all producing companies is the need to estimate the market price of crude oil. Table 10–1 shows the average deal price by year from 1995 to April of 2004. The average deal price of $4.00/boe from 1995 to 1999 increased to $5.87/boe from 2000 to 2004, a 47% increase. During these same time periods, the posted price of West Texas Intermediate grade crude increased from an average of $18.60/stock tank barrel (bbl) to $27.00/bbl, a 48% increase. The closeness of these two percentages shows the positive linkage between the movement of posted oil price and the reaction of the acquisitions and divestitures marketplace.

Most companies have a constant, yet low percentage of escalation, which stretches out for many years, built into the oil price in their strategic plans. For acquisitions, however, it is common to select a price cap for each product that represents the highest price that the company believes is reasonable, regardless of how far into the future the analysis extends. Common sense dictates that at some elevated price, fuel switching or robust exploration programs will increase the petroleum supply. This activity will eventually lower the high prices that were created by the market forces of supply and demand during the period of tight supply.

Table 10–1 Comparison of posted price versus price paid

	Assets Sold $MM	Reserves Sold MMboe	Average Deal Price $/boe
1995	3825	968	3.95
1996	3555	938	3.79
1997	9030	1985	4.55
1998	3266	838	3.90
1999	2703	860	3.14
2000	3515	883	3.98
2001	3315	405	8.19
2002	3459	631	5.48
2003	6871	1021	6.73
2004 *	648	96	6.75

Average Prices

Period	Oil ** $/bbl	Deal $/boe
1995 - 1999	18.60	4.00
2000 - 2004	27.60	5.87

* 2004 data are for first quarter only
** Approximate West Texas intermediate crude

bbl = barrels;   boe = barrels of oil equivalent;   MM = million.

Determining the Price

Reserves Growth

The concept of reserves growth should also be considered. Individuals who have spent a good portion of their careers calculating reserves in the same producing environment have observed the phenomenon that big reservoirs and fields grow more than small ones. Over time, the reserves growth observed in bigger fields is higher in percentage and volume than the growth in smaller fields. This concept should be noted when considering how aggressive an offer can be.

For the same cost, it may be more profitable in the long run to purchase a lower working interest in a large field as compared to purchasing a large working interest in a small field. There is the obvious lack of control if a small percentage and no operations involvement are acquired in a transaction, but the statistical likelihood of reserves growth in the larger property makes the lack of control an acceptable trade-off for many purchasers.

LIKE-KIND EXCHANGES   11

Property sales frequently generate capital gains. These gains will create a tax liability unless the company can offset the gain with a capital loss. If no losses are available, another way to shelter the gain is for the company to acquire similar property with the sale proceeds. This type of transaction is called a like-kind exchange.

Before taking advantage of this opportunity, there are a number of Internal Revenue Service (IRS) restrictions to consider:

- A single legal entity must sell, transfer, and receive the like-kind assets.
- The legal entity and the assets must be in the United States.
- There must be an actual exchange of property.
- Like-kind refers to the basic nature of the property, not the quality.
- Sale proceeds must be placed in an escrow account, not given to the seller.
- Once the account is funded, the company has 45 days to identify the asset to be purchased.
- Assets may be identified that exceed the value of the monies in the account.

- The qualified escrow agent must be advised of the names of the identified assets.
- The new asset must be purchased within 180 days of funding the account.
- If the exchanged properties are not the same value, a cash boot can even the deal.
- Monies left in the account after 180 days must be returned to the company, and gains tax must be paid on this amount.

The capability to defer or eliminate taxes from a sale is a powerful incentive that can enable a company to be more aggressive in an acquisition effort. The purchase price paid by a company that is using a like-kind exchange may be elevated when compared to normal expectations for a property purchase. Figure 11–1 shows that the use of a like-kind exchange can truly increase the value of a company.

Sale Only		Sale/Purchase With Like-Kind Exchange	
Sale proceeds	$3 million	Sale proceeds	$3 million
Tax basis	$1 million	Tax basis	$1 million
Capital gain	$2 million	Purchase new field	$3 million
Tax rate	35%	(Using sale proceeds)	
Tax	$0.7 million	No tax is paid	
Net cash	$2.3 million	Net property value	$3 million

In this example, the sale results in either $2.3 million in cash without the like-kind exchange or the acquisition of another property worth $3 million with a like-kind exchange. Other than the circumstance of when a company needs the cash, this like-kind exchange would always be preferred if an asset must be divested.

Fig. 11–1 Tax savings and reserves retention using a like-kind exchange

NEW FIELD DISCOVERIES   12

Characteristics

New field discoveries are unique assets in the acquisitions and divestitures marketplace. There are no production decline curves, flowing tubing pressure plots, or operating cost history to extrapolate. The evaluator does his or her best to obtain good analogies to offset fields with similar characteristics from which field performance and costs can be forecasted. At this stage in a field life cycle, there is considerable risk and uncertainty in nearly every aspect of the project. Offsetting these problems is the chance for significant upside potential that only additional time and money can identify.

The timing of first production, production rate, ultimate reserves, exploitation and development costs, and perhaps the best options for sale of the produced hydrocarbons are estimates at this early point in the life cycle. This presents the unique situation that the buyer may be as capable of quantifying these factors as the seller may be.

Reasons for Sale

It is so difficult for companies to replace production on an ongoing basis that one would wonder why an interest in any new field discovery would be sold, either partially or completely. However, it is done at times for financial reasons that may include the following:

- The project development costs are so huge that the seller does not have the investment capacity to fund its share of the future obligations, or the seller's company is not large enough to shoulder the total risk.
- The success of the drilling to date allows recovery of multiples of the sunk costs.
- The seller has other opportunities that are deemed to be a better use of cash and staff resources.

At times, there are risks that cannot be quantified or managed to the company's satisfaction, and this circumstance generally leads to a partial or complete sale of the discovery. Some examples of this risk are:

- Lack of reservoir continuity or inadequate geologic understanding of the prospect
- A belief that reservoirs will be small and require a large number of take points
- Expectation of low product prices that will render the discovery marginally profitable
- A high mechanical risk if the project needs cutting edge technology

Some reasons why companies are interested in acquiring new field discoveries despite the risks are:

- The fact that a discovery has been made greatly reduces the high-risk rank exploration aspect of the project.
- The early time flush rate of production will be captured and provide good cash flow.
- Even if there is significant upside, assets in this phase of the life cycle generally do not receive a high premium, thus purchase price multiples are reasonable.

Buyer's Evaluation

The uncertainties a buyer faces include the potential to lose the deal if a co-owner exercises a preferential right to purchase. In this situation, the co-owner has the right to match the deal that is negotiated and claim the acquisition. When operating agreements contain a preferential right-to-purchase provision, potential buyers must balance the cost of the effort to make the acquisition analysis against the likelihood that the deal would be lost in this manner.

The market multiples that are calculated from the purchase price for a new field discovery will be much less than the published multiples for producing property. This is for two reasons:

1. The investment that is required to get the field on production represents a one-time cost that will be spent as a field investment rather than as purchase cost to the seller.
2. First production from the field is in the future, thus the present worth of the reserves is less than that of the same reserves that are on stream.

Representative market sales price multiples can be calculated by adding the pending investment to the cash purchase price as the first step before the balance of the calculations are made. This will change the multiple to be a better, but not a perfect, match to producing property sale benchmarks.

Deepwater Example

Deepwater discoveries in the Gulf of Mexico are frequently targeted for ownership changes prior to field delineation, a commitment to development, or first production. A partial sale of a new field discovery is common to keep these high-cost, high-risk assets aligned with company strategy as part of a larger portfolio. Figure 12–1 gives a few of the options that are available to a company with a deepwater discovery and shows the sunk costs that were spent to date.

Sunk costs:	
Lease acquisition	$ 500,000
Seismic, geological, & geophysical	$ 1,500,000
Drill exploration well	$10,000,000
Total	$12,000,000

Fig. 12–1 Deepwater Gulf of Mexico discovery

Assume that the test well discovers 50 million boe. The project has a market value worth $25 to $50 million, using a $0.50 to $1.00 per boe rule-of-thumb value, depending on the ultimate size and production scheme chosen for full development.

- The point forward options are:
    1. Complete divestiture—Sell out: The sales price should recover double to triple the sunk costs.
    2. Partial divestiture—Sell down: New partner should pay for the next well, possibly even a flow test in the new well in exchange for a percentage of the equity in the project (one third to one half equity is typical depending on the risks and potential).
    3. Maintain current interest—By funding the next well entirely, the company retains full control, ownership, potential, and risk of the prospect. If a dry hole is drilled, the worst case would be that the prospect is not economical to develop. If the well is as expected or better, the market value of the prospect could double.

CONSTRUCTING THE OFFER 13

Options and Alternatives

Although the most debated and admittedly most important component of the offer to purchase is the cash amount, a number of other factors do weigh in on the attractiveness of the proposal. There are many incentives that a buyer can offer to encourage the seller to part with the property. In general, each will carve away some of the buyer's profit potential, but done fairly it can be a win-win situation. Some examples of what can be offered to the seller are:

- Price protection in the event product prices rise above a predetermined ceiling
- An overriding royalty interest on successful exploration activity
- Bonding protection for abandonment obligations

In many cases, the buyer has a prior relationship with the owner of the targeted asset and will discuss the subject field before making the offer in writing. At times, extenuating circumstances prevent a potential seller from divesting a property, and the phone call will save both parties considerable work. Other times, if the seller is interested, this is the opportunity to tell the buyer what terms and conditions are desired, and what conditions may be intolerable, so the buyer can consider these factors in the opening written offer. Sometimes it is difficult to move off of a position once it is in writing, and when the formal offer is preceded by verbal communication, some potential problems can be solved before they are created.

Frequently the buyer requires financing to obtain funds for the transaction. Many sellers are hesitant to begin negotiations under this premise as considerable time and effort may be expended to get to closing, and then the transaction may not close because of the lack of financing. Until the financial institution reviews the property evaluation, it is not known whether or not the agreed-upon selling price will be supported by the reserves. If a nonrefundable deposit is offered by the buyer in the original proposal, the seller gets some comfort in knowing that there is deposit money nestled in his account if the deal does not close as a result of a financing problem.

Deposits are also requested when there is some doubt on the seller's part that the buyer can close the transaction. If the buyer offers a nonrefundable deposit to purchase a property and lists the situations under which the deposit will be forfeited, the sincerity of the buyer is noted as one who will not walk from the deal too quickly.

Many property sales involve fields that are in the mature period of their life cycle. Buyers who intend to resurrect this type of property and transform it into a healthy asset are prepared to invest considerable human and capital resources into the effort. The acquisition and the subsequent redevelopment plan each have a fairly high element of risk.

For this reason, if there are co-owners in the targeted field, the prospective buyer may state in the offer that all co-owner interests must also be acquired for the deal to close. If 100% of the property can't be acquired, the buyer will pass on all of it. The buyer:

- Wants to control the timing and scope of the redevelopment program
- Wants to spread resources and overhead over the entire field to keep unit costs down

- Does not want to carry co-owners who are unwilling to fund proposed investments

If the buyer is successful in revitalizing the newly acquired property, it can follow up the purchase with an attempt to acquire nearby assets that are similar. The buyer would try to duplicate the revitalization effort profitably and at lower risk if proprietary technology or information was employed successfully.

Appraised Value

Many of the processes used to buy and sell property are quite flexible and dependent on the situation at hand, but the appraisal of property should be consistent from one analysis to the next. A standard methodology incorporating accepted reserves definitions, consistent risking of reserves by category, and a cost analysis based on past experience and expectations is necessary.

Product price and escalation rates are analysis inputs that are less science-based. These factors represent the best guess of the analysts who have the responsibility to make those forecasts.

The risking of reserves combines technical know-how and experience. For starters, the reserve calculations need to be done using standard SEC reserve category definitions. This ensures consistency from one evaluation to the next, consistency within the industry, and common ground when communicating with lenders and consultants.

Reserve adjustment factors were provided by a cross-section of the industry to the SPEE in a survey taken in early 2003 (see Table 13–1). This collection of industry data, entitled *Survey of Economic Parameters Used in Property Evaluation*, has been published annually for 22 years. The survey gathers evaluation parameters from producers, consultants, bankers, and government sources. The perceived risk of these reserve classes on a comparative basis has been relatively consistent from year to year.

Table 13–1 Reserves adjustment factors

	Acquisitions	Loans
Proved producing	100%	100%
Proved shut in	85%	78%
Proved behind pipe	75%	75%
Proved undeveloped	50%	50%
Probable producing	34%	0%
Probable behind pipe	25%	0%
Probable undeveloped	20%	0%
Possible producing	3%	0%
Possible behind pipe	0%	0%
Possible undeveloped	0%	0%

* Median figures

The table shows the difference in conservatism between an acquiring company and a lender. Once this gap is understood, it is much easier for sellers and buyers to work together to close a deal when the funds for the purchase need to be borrowed. As shown in the table, probable reserves generally do not support loan value, thus purchasers will diligently work the data that backs these locations with the hope of upgrading the associated reserves to proved status.

The SPEE survey also shows that the reserve adjustment factors are applied to the discounted cash flow that is derived from a forecast of unrisked production approximately 52% of the time, and that the factors are applied to the reserves prior to scheduling production approximately 48% of the time. The variability in the approaches to the analyses further supports the recommendation that consistency from one transaction to the next is important in a company's acquisition and divestiture evaluation efforts.

Different risk factors are placed on the reserves categories depending on the purpose of the analysis. Figure 13–1 shows the difference in risk that would be applied to the reserves volumes when the analysis is being done by an evaluator for a selling company versus an analysis that is done by a bank for the purpose of determining whether a loan should be granted to a purchaser.

The most commonly used evaluation method to determine property value is to calculate the discounted cash flow, which was used by 86% of the SPEE survey respondents. The average discount rate that was used to determine the present value of a future cash flow stream was 10.7%.

The annual SPEE survey results address many critical parameters not mentioned in the previous discussion. They include product pricing escalation, operating and drilling cost, inflation, and opinions on discount and borrowing rates. Because most evaluations

	Unrisked	SPEE survey		Loan	
	Mboe	Risk %	Mboe	Risk %	Mboe
Proved production	2000	100%	2000	100%	2000
Proved shut-in	1000	85%	850	78%	780
Proved undeveloped	800	50%	400	50%	400
Probable production	2000	34%	680	0%	0
Totals	5800		3930		3180
% of total			68%		55%

Mboe = thousand barrels of oil equivalent; SPEE = Society of Petroleum Evaluation Engineers

In this example, there are four classes of reserves in the analysis. The reduction in reserves because of risking from the view of the seller to the bank for the buyer is 750 Mboe. If the seller requires $3/boe for the high-risk reserves that the bank will not use as collateral, then a shortfall of 750 Mboe x $3 = $2.25 million exists that the buyer needs to have at closing, above the loan amount, to make the acquisition.

Fig. 13–1 Reserves categories–different risk perception

in the industry are either not made in support of transactions or are statistically not successful in making an acquisition, one could assume that to be successful in the acquisitions marketplace the parameters used in a company analysis must be more aggressive than the survey results. For this reason, the survey parameters can probably be considered to be a good benchmark for calculating the fair market value of a property, excluding its strategic or investment value to the successful bidder.

The discount rate that is used in the evaluation is very important. This figure has a great impact on determining the amount of the opening and highest bids. Each company has proprietary reasons that guide its rate decision. For some it is the cost of capital; for others it is the return from the least profitable investment opportunity in the portfolio. For publicly traded companies it may be the figure that represents the borrowed cost of money plus the cost of paying dividends. In summary, the lower the discount rate used in the analysis, the more a company can pay. Thus, the company with the lowest discount rate and a competitive analysis relative to other potential purchasers will likely be competitive in an acquisition attempt.

When a seller markets assets that are close to depletion, there is generally more certainty in the reserve calculation and the major inputs in the analysis. For these sales, sellers may decrease the discount rate used in the internal retention analysis to reflect the comparatively lower uncertainty in these properties relative to the remainder of the portfolio. This will elevate the minimum acceptable selling price above what it would be if the standard company discount rate were used.

Strategy of Amount

There are a number of strategies for picking the amount of the opening offer. Little public proof exists to show which approach works best, so most offers are based on experience and intuition.

The only situation that appears to universally get a negative reaction is the low-ball offer. This is when the buyer offers a price that is significantly below the fair market value and in return expects a serious counter offer. Because the offer is not made in good faith, the seller rarely counters and has no motivation to engage in negotiations with the potential buyer. If the seller did counter, it would place a cap on what the field would be sold for; the seller has no need to do this. For this buyer to be taken seriously and get the seller to engage, the follow-up offer, if there is one, needs to be very strong.

The objective for the buyer is to make the lowest possible opening offer that arouses the seller enough that a counter offer is obtained. Given the interpretive nature of evaluation science, this is not a definitive calculation and the fear of the serious buyer is that the seller may interpret the offer as a low-ball bid. For this reason, a slight bump to the figure determined by using this logic is usually a good opening offer. With this approach, the buyer always assumes there will be a need to increase the offer in order to close the deal, but does not know how much of an increase will be needed at the outset.

When both parties are serious but they cannot get together on price, it is common for the buyer to ask what price is needed to make the deal and wrap up the negotiations. If the seller really does want to make the deal, the price may be disclosed. Frequently, however, this approach is interpreted as a negotiating tactic on behalf of the buyer in which the seller is pressured to disclose his minimum acceptable price. If the buyer is successful in getting the seller to provide a price, the buyer does not pay any more than is necessary for the acquisition.

The opening offer is made with the assumption that some improvement in price will be required during the negotiating process. The buyer always allows some room to increase from the first offer, knowing that the first offer does not represent the highest offer that can be made.

The opposite situation can occur when a buyer consciously submits a conditional offer that is a very high offer with no intentions of closing at that amount. A high offer always gets the attention of management and usually results in quickly engaging the company that made the offer to close the deal as rapidly as possible. Under this scenario, the buyer will then note reasons to lower his offer after the seller has engaged in negotiations. This generally upsets the selling company, which would not have begun negotiations if it had known the offer was going to be reduced. Offers that start with this plan rarely result in a closed transaction.

The situation discussed previously occurs when a company is making an unsolicited offer. When the buyer is at an auction or bidding on a property through a broker, the approach is different because of the competitive nature and limited timeframe of this marketing media.

When a broker is marketing the property, the seller hopes to get the *outlier* bid—the bid that is considerably higher than the field value or the cluster of value around which the majority of bids are made. When an outlier bid is received, the seller generally thinks the buyer has made a mistake in the evaluation. In reality, at times the company making this high bid knows of upside potential that is unknown to the seller or the competition, and the buyer has actually made a prudent offer that will result in a profitable transaction. When brokers sell property, there is an element of mystique to the process, because the potential buyers who are engaged do not know which other companies are in the fray or how many were contacted initially.

Sales Package Examples

The following discussion on three particular sales packages shows the spread of offers that were received several years ago in brokered deals for smaller fields in Texas, Louisiana, and Mississippi. To anyone who has not marketed property in this manner, the broad range of the offers, combined with the large variation from appraised internal company value, are quite surprising. The purchasers were each given the same data set and had the same length of time to evaluate the information. In each of the three packages, the buyers had huge differences in value assigned to the fields.

The Texas sales package is presented on Figures 13–2 and 13–3. The data show the range of offers on four of the six fields that were in the package. Brochures were sent to 32 interested companies, of which 16 made offers. In all, 57 offers were made on the fields,

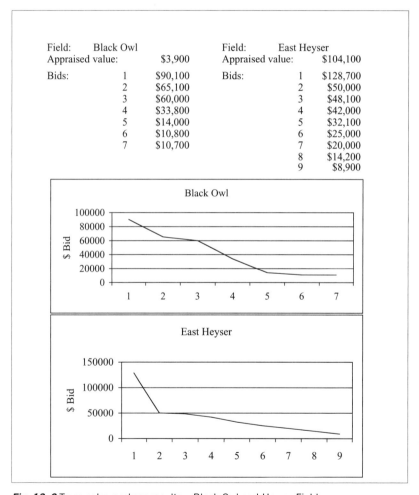

Fig. 13–2 Texas sales package results – Black Owl and Heyser Fields

which were marketed as individual properties. A comparison of the offers to the seller's calculated value shows a disparity that is nearly unbelievable, both at the high and low ends of the spectrum. The data given to the buyers would seem to be from a different data set than that which the seller used for the internal evaluation.

Figure 13–4 shows the Louisiana sales package. This package of six fields was originally marketed as a single sale in which the appraised value exceeded $1,000,000. Only two offers were received,

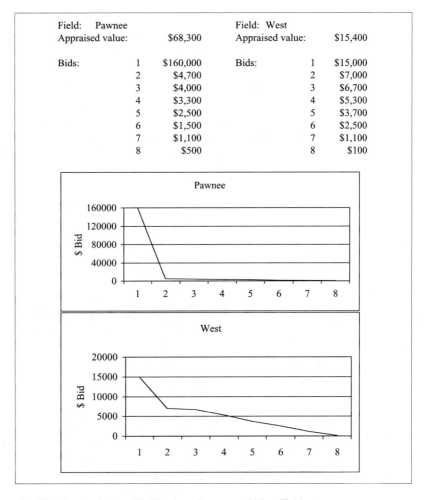

Fig. 13–3 Texas sales package results – Pawnee and West Fields

both in the range of $750,000. Because the bids were too low to make the sale and buyer interest was minimal, the package was split into individual field sales in an attempt to interest the market and increase the offers to an acceptable level. As a result of the new effort, 11 companies submitted offers and most fields sold at a combined price of more than $1.2 million. The range of offers was tighter and more reasonable than was experienced in the Texas package, but not all of the fields sold. Figure 13–4 shows that the Fordoche field, although it received good interest, did not receive an acceptable offer until negotiations with the high bidder were successful.

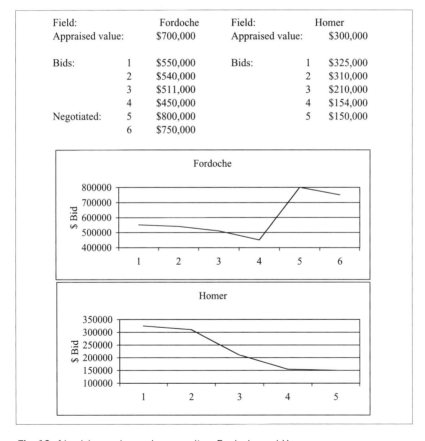

Fig. 13–4 Louisiana sales package results – Fordoche and Homer

In the third package, 14 fields in Mississippi were marketed individually. Fifty-four brochures were mailed; 29 companies responded with offers, for a total of 67 field bids. The range of offers by field was extreme, as was generally the difference between the internal appraised value and the high offer. Figures 13–5 and 13–6 show four of the fields. Attempts to discern what caused the variation from appraised value were usually not successful. However, many of the fields were nonoperated and were not worked very hard by the seller while the field was in the portfolio, leading one to assume that the property's retention value was underestimated.

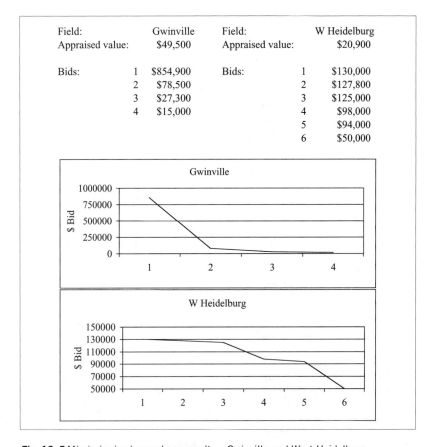

Fig. 13–5 Mississippi sales package results – Gwinville and West Heidelburg

The appraised value for the reserves at the Gwinville field was $49,500. The high offer was $854,900, a figure that had no obvious justification, particularly when the next highest offer was $78,500. The explanation provided by the broker was that he was able to get the buyer to value the asset at prime acreage cost, because it was a large tract and offset deeper play drilling had been successful. This was a remarkable example of when a broker knew the area far better than the seller and was able to use his information to increase the sales price.

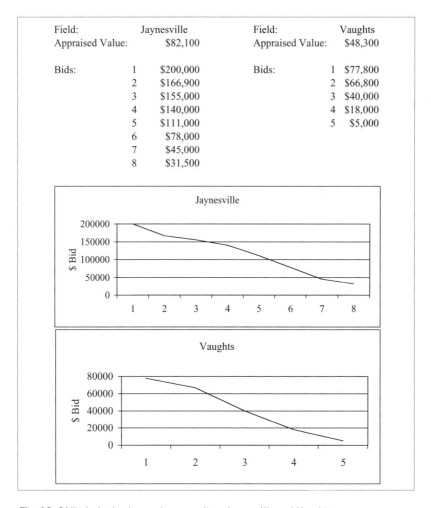

Fig. 13–6 Mississippi sales package results – Jaynesville and Vaughts

Constructing the Offer

The offer distributions for the Jaynesville and Vaughts fields are both very wide, indicating that there were probably low-ball offers as well as offers with high strategic value. Both fields were sold for prices much higher than the retention value.

From the buyer's perspective in a brokered sale, there is pressure to be as aggressive as possible from the beginning as there is generally no second chance. Acquisitions are so difficult to make and require so much effort that no company wants to lose a transaction by a slim margin. As a result, participants in offerings that are handled by brokers usually stretch their limits to the maximum. Indeed, many of the publicly reported transactions that were closed at high value multiples were managed by brokers.

PREFERENTIAL RIGHTS 14

Certain properties have preferential rights provisions stating that when one of the owners divests title to the property, the other owners have the right to take the transaction if the terms of the deal are matched. This applies regardless of the marketing strategy that may have been employed.

If the preferential right is exercised, the original company that worked hard to acquire the property loses the transaction and ends up with nothing to show for the effort. This is a costly and demoralizing blow to the would-be acquirer, but it is nevertheless industry practice to write operating agreements with this provision.

The path that an exercise of a preferential right takes is not the same for all transactions. Although the terms of the lease agreement govern the process, many times it is still interpretive from the time of notice of the sale until the closing of the transaction. In some cases the seller offers the preferential right to the co-owners with only a price, followed by a negotiation of the purchase and sale agreement with the ultimate buyer. Other times, the seller offers the preferential right after the price and contract terms have been negotiated, and the co-owners must accept the contract as it is written as well as the price.

If the seller allows a co-owner to renegotiate the contract, the first buyer can claim that the deal has changed and the original deal should be allowed to close. Each party must be careful to observe the specifics of the operating agreement in this regard to avoid such problems. There is always a prescribed time limit in the agreement for the co-owners to respond to receipt of notice. Typically the response period is 15 to 60 days.

If a package of property is sold in a single transaction, and one or more of the properties has a preferential right, the buyer must allocate a portion of the total price to the properties that have the right. Although allocations are already done for tax or other strategic reasons, an allocation for preferential rights takes on more significance because the loss of the field from the package is a potential outcome.

The price allocation is frequently done in a manner that purposely skews the value of key fields higher than their evaluated worth. This is done so that if the co-owners exercise on a key field, the buyer profits immediately simply by working the allocation process. The other reason the buyer may allocate high on the key field in a package is to try to ensure the field will not be lost to a co-owner via the preferential rights process.

If the majority of fields in a package have preferential rights, and too many of them are exercised, the package that remains for the original buyer does not satisfy the objectives of the original buyer's initial offer. Both parties understand that this situation can occur, and it must be addressed in the purchase and sale agreement before the preferential right notices are sent to the co-owners. Normally a percentage, typically 50%, is chosen, which indicates that if more than that portion of the package value is lost to the co-owners, the original buyer is not required to close the transaction on the balance of the package. Obviously, however, the buyer would always have the right to close on whatever remained in the package after preferential right elections, regardless of how many fields were lost.

If more than the specified limit of property value is taken by preferential rights and the buyer does back out of the deal, the seller may be forced to retain the lower quality fields and sell the better fields—a result that is opposite of the seller's original intentions. For this reason, the seller inserts a provision in the purchase and sale agreement stating that if this situation occurs, all of the fields in the package can be withdrawn from sale and none of the fields will be sold to co-owners.

The seller does not get involved in the allocation process for the following reasons:

- The seller is exposed to litigation if the co-owners claim an improper allocation was made.
- The allocation generally does not impact the seller's business.

The one situation where the seller may get involved in the process (but not the price) is when the buyer wants to close the transaction quickly. The seller is asked to contact the co-owners and request that they waive their preferential rights earlier than the time period allows. This may be requested when the buyer's financing is available for a limited time. Most sellers will comply with this request as a fast closing is generally in their best interest also. When a valuable property is involved, it is unlikely that the co-owners will accelerate their review period and grant the waiver; however, as there is no benefit for them to do so.

There is a situation in which the preferential right provisions may not apply and the co-owners of divested properties do not have the right to match the deal. When a whole company is sold or merged, or when a whole subsidiary or operating unit of a company is divested as one entity, the preferential rights generally cannot be exercised. The logic for including this exemption language in the operating agreement is that the value of a corporate entity or operating unit is much more than that of the combined value of the individual assets comprising it. Thus, it would be difficult to allocate value under these circumstances, and the company making the acquisition would lose more than the value of the asset if a preferential right were exercised.

Preferential Rights

BONDING PROTECTION 15

Before selling property to any company, the seller needs certain assurances that the buyer is a competent producer and will handle the end-of-life costs in a lawful manner. Some information describing the current status of the company making the acquisition should be reviewed, such as current market capitalization, reserves volume, production rate, and abandonment obligations. Financial statements are also needed. If this review indicates that the buyer does not have the financial capacity to make the purchase, then an additional layer of protection is needed.

With the sale of certain large or old fields in particular the costs of future abandonment and environmental obligations can be significant. A lease usually requires that the lessee return the property to its preexploration condition upon cessation of production.

At some point, the removal costs of facilities and pipelines, combined with the well abandonment, can be greater than the value of the reserves. For old fields, the remediation costs to return the land surrounding the pits, spills, or tank batteries to original conditions can be staggering. In many cases, fields in an offering have negative characteristics that represent a particularly large risk for both seller and buyer. Sellers lessen this risk at times by spending the money and performing abandonment and remediation activity prior to the marketing effort. This lowers the cost and risk to the buyer, who pays a higher purchase price up front to acquire a property with reduced end-of-life costs.

The sellers of producing property do not want to sell a field and have these costs returned to them by a regulatory agency in the future if the buyer does not perform abandonment activities or remedy environmental damage properly.

Larger companies will require protection that guarantees that the funds to pay for these costs will be available in the future, as shown in Figure 15–1. If the buyer has a market capitalization that is deemed low relative to the costs, the seller will require bonding protection to insure that the monies will be available when needed. This takes the form of a performance bond that is purchased by the seller from a bonding agency. The buyer pays annual premiums to maintain the coverage. At times, for additional seller protection, insurance is purchased before the closing guaranteeing that the future premiums will be paid to avoid a lapse in the policy.

One must do the homework to be reasonably certain that the company issuing the bonds has the financial strength to perform if a number of the bonds issued by the company default at the same time. It is a step in the due-diligence process that will be worth the effort if at some point in the future the bonds need to be redeemed to pay for the abandonment or remediation costs.

If the fields that are sold are marketed while the value is high relative to these costs, a company with a high market capitalization is more likely to be the successful buyer, and the effort associated with getting bonding protection will not be necessary.

An offshore package is sold to a small company that needs bonding to ensure the abandonment obligations for the fields will be properly handled. The projected schedule of abandonment cost in escalated dollars, in millions, is as follows:

	2004	2005	2006	2007	2008	2009	2010	Field Totals
Field A	0	0	0	2.5	0	3.0	0	5.5
Field B	0	2.5	2.0	0	0	0	0	4.5
Field C	0	0	0	0	0	1.0	3.5	4.5
Field D	0	0	0	0	6.0	2.0	0	8.0
Annual Totals	0	2.5	2.0	2.5	6.0	6.0	3.5	22.5

The seller will negotiate initially to have the buyer bond the full $22.5 million of abandonment cost. As the fields are abandoned in the future, the bonded amount would be reduced to reflect the expenditures, after the Minerals Management Service (MMS) has approved the results of the operations. If the total of $22.5 million is a deal breaker because of the cost of the premiums, two options can be considered:

1. Field B would not be bonded because it must be abandoned first. The buyer would agree that Field B would be abandoned properly to get any bond reductions from Fields A, C, and D, which will be abandoned in 2007 and beyond.

2. The cost of platform abandonment alone would be bonded for all of the fields. Well plugging would not be included in the bond coverage, as the wells must be plugged prior to platform removal. The buyer could not get a reduction in bond unless the fields in total were abandoned.

Fig. 15–1 Determining bonding requirement for Gulf of Mexico purchase

NEGOTIATING THE AGREEMENT 16

Terms and Conditions

In competitive bid situations, the high bidder is usually the successful company. At times a company with a lower bid is chosen if the liabilities associated with the property are significant and a bidder with *deeper pockets* than the high bidder is deemed a better purchaser. If the gap between the high bid and the next bid is large relative to the liabilities, the high bid will be accepted.

Once a purchaser is chosen using the competitive bid process or the negotiated approach, the purchase and sale agreement is negotiated. This document establishes the baseline, or benchmark, for expectations by both buyer and seller as it relates to the properties and the performance of each company throughout the closing process.

The negotiations include such topics as:

- The indemnities that the seller wants the buyer to accept
- Protection for the specific performance of abandonment operations and surface cleanup by bonding or letters of credit

- Title defects, ownership differences, or mortgages not disclosed
- Environmental claims and identification of material conditions after closing
- Marketing restrictions and the release of the call on production
- Settlement of gas imbalances and the price per Mcf to be paid
- Excluded assets (i.e., vehicles, boats, communications equipment, warehouse stock)
- Excluded obligations such as prior personal injury or property damage claims
- The amount of the performance deposit and when it is to be paid
- Conditions under which the buyer can back out without forfeiting the deposit
- The date by which the transaction will close and repercussions if it does not close
- Assistance rendered to the buyer to obtain operations, if desired
- The period of time and breadth of investigation to be allowed in the due-diligence process
- The assistance to be given by the seller to the buyer to obtain assignments from landowners
- Whether originals or copies of files and other records will be transferred to the buyer
- The content of press releases and which company can make public notification

All negotiators are not created equal, and it is wise to have experienced negotiators on the team. The need for legal representation is obvious, but evaluation engineering expertise is also beneficial to properly assess the value of each term and condition in the agreement.

In chapter 23, Lessons Learned, numerous negotiating tips are provided that will save money and time for all parties in a transaction. Failure to follow some of the recommendations can injure a company's reputation and prove to be costly.

Effective Date

The seller will decide on the effective date of the offering and communicate it to the marketplace in the initial solicitation to the potential buyers. This is the date:

- Before which all revenues and costs are retained by the seller
- After which all revenues and costs flow to the buyer

The effective date is chosen carefully based on a number of factors, including:

- Whether any well work is in progress, the risk level of the work, and when it is expected to be complete. If the well work is high risk, the seller might pick the sale effective date before completion of the work; if the well work is low risk, the effective date may be picked after the well should be on stream.
- How current the seller wants the data in the data room to be. An effective date that is chosen before the opening of the data room will give the buyers accurate historical data up to the effective date to use in their evaluations.

- The length of the time lag between the effective date and the closing date. A short time lag lessens the likelihood that well performance can change, which could result in a renegotiation or possible loss of the deal.

- The speed in which the company can mobilize and conduct the sales program once it decides to sell the property. A date in the near term places a sense of urgency toward accomplishing the sale.

At times, particularly in situations when negotiations are difficult or due diligence is complex, the closing process becomes much longer than either party originally expected. The effective date may be pushed out further in time, lowering the purchase price by the amount of the interim cash flow. The lower price may make it easier for the purchaser to close the deal, thus the date shift benefits both parties.

A shift of effective date backwards may be suggested by the buyer if the sale price includes consideration of a significant amount of unproved upside potential. If the buyer needs financing to close the transaction, the effective date can be shifted backwards so a higher percentage of the value of the deal is in the proved reserve category. This reduces the risk to the lender, enabling the buyer to obtain the loan, and the seller simply exchanges cash flow for a higher purchase price. Because cash flow is taxed at the corporate tax rate and the sale price may be subject to capital gains and losses (at a lower tax bracket), this shift may result in a more favorable tax impact to the seller.

DUE DILIGENCE 17

All buyers conduct due diligence, which is an investigation of field conditions as well as records kept in the field, the office, and by regulatory and government agencies. This is the stage of the process in which a buyer can discover negative information that may be so distressing that the company can back out of the transaction or lower the purchase price.

A field trip is conducted to review, at a minimum:

- That the wells are producing the reported volumes without any hazardous attributes
- That the facilities are in good condition and contain the equipment as reported
- That the capacities of flow lines and gathering lines or compression capacity are adequate
- That the environmental status and abandonment costs of the property are within acceptable limits

The review of records should include:

- Land and lease reports
- Well bore schematics and facility design

- Production reports
- Regulatory and governmental reports
- Product marketing options
- Well and field files
- Legal files

The due-diligence process is very important as it is the buyer's one chance to identify information that was not previously disclosed. When negative information is discovered, a reduction in price or change in terms as compensation is usually requested. If a severe problem is noted that cannot be cured, the buyer may decide to back out of the transaction.

At times a company will request a field trip to conduct due diligence during the bid preparation period. This request is generally refused by the seller for several reasons:

- If all the companies that were interested in the field requested field visits, it would be very difficult to schedule them all.
- One of the companies visiting the field may discover something that the seller intended to show the successful bidder only.
- Industry practice is to wait until the successful bidder is chosen and have just one company make the field trip.

An excellent checklist of information to use as a guide for the review of a seller's accounting records is the booklet published in 1995 by COPAS entitled *Property Acquisition Checklist*. This booklet contains a well-organized, exhaustive list of data that should be reviewed by the buyer to ensure there is a complete understanding of the asset while the acquired property is absorbed into the company's operations. It begins with a preliminary acquisitions analysis and progresses through the various evaluation and due-diligence phases, culminating in a review of the agreements for purchase and sale and the handling of expenses, revenues, and royalties.

Some buyers approach the due-diligence period very aggressively, knowing that it is the only chance to reduce the original offer. As a result, those on the due-diligence team may be compensated with a percentage of the amount of the sale price reduction that they manage to get the seller to make. For the seller and buyer, the due-diligence period is one during which they must be cautious and knowledgeable about their rights under the purchase and sale agreement.

GOVERNMENT APPROVALS 18

To be finalized, some transactions require the approval of government agencies. This step may be at the federal, state, and/or local levels.

Very large transactions, more than $100 million, must pass a review by the Federal Trade Commission (FTC) to protect the American public against the loss of competition or creation of a monopoly in the marketplace. This review process, which is done in compliance with the Hart-Scott-Rodino Act, can take several months.

A "consent to assign" provision is common in many leases. This term gives the landowner the right to block the transfer of title from the seller to the buyer if there is good reason to do so, or to ask for changes in the purchase and sale agreement.

For offshore Gulf of Mexico sales, the Minerals Management Service (MMS) branch of the federal government will review the fiscal position of the buyer relative to the ongoing responsibilities that are associated with the property, as well as the ultimate abandonment cost. The agency may require bonding or a letter of credit to assure prudent operations, payment of royalties, and final restoration of the lease to predrill conditions.

At the state level, the government may own the water bottoms of lakes, rivers, and bayous. When state leases change hands, the state's approval is also needed. In the marsh areas, huge abandonment and cleanup costs are forecasted, and the government is very attentive to be certain that these fields are only sold to qualified buyers who are capable of funding these projects when the time comes.

CORPORATE SALES PROGRAMS 19

From time to time, most large companies find it is appropriate to divest large blocks of property to satisfy corporate objectives. These objectives ordinarily signal a new direction or focus for a company, initiated by new management or forced by the opinions of the analyst community.

The story behind the motive for a *portfolio adjustment* program is always well thought out and painted with a brush that emphasizes the long-term health and competitiveness of the company. History has shown, however, that although the desired parameters are often improved by the results of the sale program, other financial or operating parameters that may be out of vogue at the time of the sale are often harmed. It may take several years for this negative impact to be noticed, and generally there is no resolution to the loss after discovery.

Examples of this effect are as follows:

- When a sale of fields having *low rates of production* are targeted to elevate the average production per field in the company portfolio, it may result in selling fields with low lifting costs, high unit profit margin, or exploration potential.

- When a sale of fields having *UOP write-off rates and/or high lifting costs* are targeted to elevate the average earnings per unit of production, it may result in selling fields with exploratory potential or good production rates.
- When a sale of fields having *low working interests* are targeted to get more focus and impact from the technical staff, it may result in selling fields with high production rates, solid exploration potential, or high unit profitability.

Undoubtedly, some of the fields that were divested in each of these programs are actually the type of field that the company will seek to acquire in the future. It is an unfortunate and sometimes embarrassing situation. The common cause is the need to balance the long-term view versus the need to satisfy analyst short-term opinion. In the balancing act, some fields are sold that should have been retained. More than a few companies realize after a sale program that fields that should have been retained were divested, caught up in a metrics sort that did not place enough emphasis on long-term value. The net result is that the company has a portfolio that is smaller and lacks sufficient exploration potential.

Three large U.S. sales packages were sold by a major over a 10-year period. The packages were not coordinated with each other, because each sales program was championed by a different management team. Over the long haul, the asset base deteriorated and good fields were sold, harming the portfolio beyond expectations. The sequence of the sales were as follows:

- Program #1, 1989—After making a costly cash acquisition a few years earlier, funds were needed; the large fields that were burdened by a Purchase Cost Adjustment in the UOP rate were targeted because of their low earnings rate per barrel. Many fields that were worth more than $10 million each were sold. *Volume was lost that was actually the target of the original acquisition.*

- Program #2, 1994—A large U.S. sales program captured all fields with a high lifting cost in an effort to make this parameter more competitive in benchmarking comparisons by Wall Street. Most of the smaller fields in the portfolio were sold. *Many of the fields had upside potential not fully studied using 3D seismic; other fields were very profitable, but were just too small to be retained.*

- Program #3, 2000—The worldwide portfolio was reviewed and fields of all types were targeted for sale, principally those with low producing rates, low earnings, and high abandonment or environmental costs. *The selling price captured a lower-than-desired multiple for nonproved reserves, and the remaining asset base was found to have fewer-than-expected investment opportunities for the future.*

The moral of this story is to be very careful when establishing the selection criteria when selling packages of fields. Look at your company's objectives closely, and look beyond the parameters that are specific to that sales program at that time. Be certain that selling the chosen fields does not benefit short-term metrics at the expense of long-term company performance.

INDUSTRY ACTIVITY 20

Historical Acquisition Prices

It is always helpful to be aware of the competition in any endeavor. The United States is a very active producing-property A&D market. Many companies track various aspects of this market. The Scotia Group, Inc. Web site, for example, contains a substantial sampling of this domestic A&D information beginning in 1979. The database is sorted by month and includes buyer and seller names, transaction prices, reserves volumes, and production volumes. Plots that show trends in the industry are also provided.

Figure 20–1 illustrates the immense industry effort that is placed on the evaluation of acquisition targets. The data show that more than $600 billion of publicly reported acquisitions have occurred during the past 20 years, most of it since 1998 when the major companies began the run of merger activity. Because the majors are all international in scope, reserve and production data are not shown in the data table as the figures are intended to be domestic and not international. When one factors in the statistic that many deals are

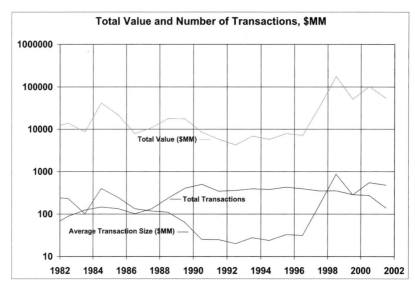

Fig. 20–1 Total U.S. Transactions Reported

worked in vain for each one that is closed successfully, it is not difficult to project that more than $1 trillion of producing assets were evaluated during the past two decades. This exhibit also shows that more than 6,000 transactions occurred during this time frame. With numbers this large indicating the resources that are devoted to this aspect of the business, it is obviously necessary for each company, large or small, to be as expert as possible in A&D activity.

The price that is paid for reserves is potentially the most competitive parameter that is tracked. Figure 20–2 shows the price paid for the past 20 years for reserves by British thermal unit (BTU) equivalent and by price equivalent. This graph shows that for the last six years on the plot the price of energy has risen (absent inflation), measured on a BTU basis. The increase in price paid for natural gas relative to oil has been obvious in the marketplace recently and has had a great influence on this trend. Another, more subtle reason for this price rise is the inclusion of nonproved reserves in the buyer analysis. The Scotia Group has noted in recent publications that the inclusion of lower-confidence reserves and the use of technological advances are two approaches that successful buyers have adopted,

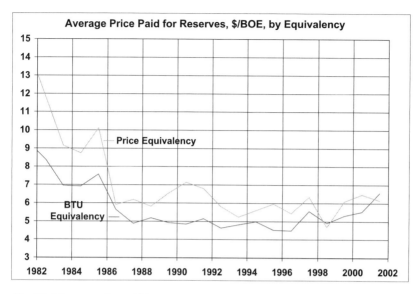

Fig. 20–2 Average price paid for reserves

and that this trend is not likely to ever reverse itself. Much of this unproved reserves volume is not reported in post-transaction statistics, causing the price paid per BTU to rise.

Figure 20–3 shows that despite the huge variation in annual wellhead prices that has been experienced year to year, the average price paid for reserves from 1986 through 2002 varied less than $2/boe! This remarkable statistic shows that buyers do not factor the complete value impact of near-term price anomalies (high or low) into their price deck forecasts. However, when prices are high, sellers become scarce because they will not willingly sell into a market that does not pay for full value of current production.

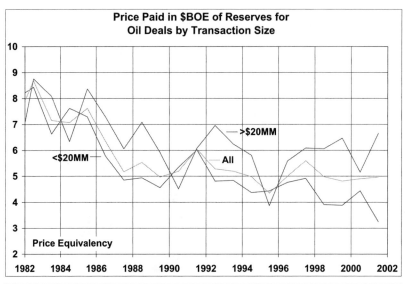
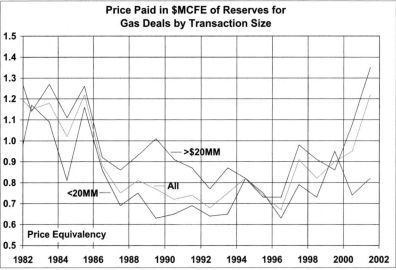

Fig. 20–3 Total value and number of transactions

Justification of Elevated Multiples

By mid-year 2000, it was apparent that an increase to the historical price paid per boe was not a short-term event. Numerous consultants and bankers who work on acquisitions regularly were contacted and asked for their viewpoints of the measures that were being taken by successful buyers in the marketplace at that time. Their assessments had one common thread—aggressiveness in selected evaluation parameters leads to the payment of high premiums relative to past market multiples. Companies had to do one or more of the following to be the high bidder:

- Push the envelope of product price, reserves volume, and reserves risk to uncomfortable levels; place no risk on proved reserves; and place value on high-risk potential, possible reserves, and exploration drilling

- Conduct an intensive mapping effort to gain an exceptional knowledge of field geology

- Identify a unique factor in the asset to increase the value of the asset over what the competition could create

- Show genuine corporate energy toward the targeted assets and involve all company resources in the effort

- Expect profit from the acquisition to be derived from the upside potential, not from the proved reserves

- Use low discount rates and exclude corporate overhead costs from the analysis

Many of the previous measures would not ever be taken by the majors in the industry. Given this and the low likelihood of success that would be indicated because of the lack of perceived aggressiveness, many insiders of the majors believe that it is a waste of time to participate in competitive bid situations. For this reason and the belief that the exploration programs will discover sufficient new reserves, active acquisition groups are not as common in major companies as they are in smaller firms that may depend on acquisitions to add reserves.

Other comments made by the consultants were that deals that were not hedged would never pay out if product prices did not remain high, and that SEC reserve definitions were being pushed beyond intended limits. Given the upward trend in product price since this informal survey in 2000, the buyers who approached their analysis in an aggressive manner were probably rewarded for the risks that were taken.

COMPANY MERGERS 21

The industry has experienced a number of major company mergers (Exxon-Mobil, BP-Amoco-Arco, Chevron-Texaco), as well as others that may be smaller but still quite significant, that have reserves multiples far in excess of historical asset-based transaction multiples. There are three reasons for the high acquisition prices:

1. *Pooling* accounting allows a company to purchase (or merge with) another company using stock as the currency. With this method of accounting, a fractional number of the acquired company's shares are determined to be equal to one of the acquiring company's shares—the acquiring company simply absorbs the stock and existing tax and financial accounting basis for the acquired company's assets. This is an avenue for the acquiring company to pay an exceptionally high price for another company with no obvious, immediate negative financial ramifications, because the excess cost is not distributed to the assets as it is done in purchase cost accounting. Dilution of the acquiring company stock for the stockholders is the downside for pooled transactions in which a price in excess of appraised value is paid.

2. The integrated majors have a large amount of value associated with nonproducing assets (e.g., foreign exploration programs, mid-stream activities, and refining and marketing operations) and with research and technology efforts to identify fuel sources that may be economical in the distant future. These assets will elevate the price paid per boe of reserves.

3. An evaluation technique that is not used for asset evaluations is that of calculating and reinvesting the *free cash flow* that an entity will generate. The profit that the targeted company is expected to generate is invested in the analysis into projects that are currently unfunded in the acquiring company's portfolio. This methodology gives the true upside value of an acquisition, although it leaves no room for error if the profits do not materialize.

Only time will tell on a case-by-case basis if the value of the added nonproducing assets in Reason 2 outweighs the effects of Reason 1 to justify the high reported multiple figures. Several of the companies that track and report the cost per barrel of U.S. transactions for the industry include mega-mergers that used pooling accounting in their statistics. As a result, their reported multiples for the past few years are nearly useless as predictive indexes for asset-based transactions. The high multiples, combined with the multibillion dollar cost of these deals, eclipse the figures that are reported for asset-based transactions, thus one does not know what the reserves actually sold for when the databases including mega-mergers are used.

Table 21–1 shows the historical market multiples of cash-based acquisitions versus stock-based mergers since 1979. The data show the huge premium paid for reserves in mergers—$10.33/boe versus $4.78 paid in cash deals. The premium is also reflected in the price paid for production in mergers—$32,770/boepd versus $12,980 paid in cash deals. It would be extraordinary if the reserves acquired in the mergers were actually worth a $5.55/boe premium, because:

Table 21–1 Relative cost of cash versus stock-based transactions

	Cash Based Acquisitions	Stock Based Mergers	Total Database
Reserves data available:			
Number of deals	1,609	128	1,737
Total acquisition cost - $MM	76,587	358,665	435,252
Total reserves - MMboe	16,033	34,714	50,747
Average size deal - MMboe	10.0	271.2	29.2
Average cost - $/boe	4.78	10.33	8.58
Production data available:			
Number of deals	954	68	1,022
Total acquisition cost - $MM	58,996	174,703	233,699
Total production - Mboepd	4,545	5,331	9,876
Average size deal - boepd	4,764	78,400	10
Average cost - $/boepd	12,980	32,770	23,663

Data based on transactions from 1979 through 1st Qtr 2004

Boe = barrels of oil equivalent; boepd = barrels of oil equivalent per day; Mboe = thousand barrels of oil equivalent; MM = million.

- Operating cost savings and synergies are at best $2/boe in fields that have a strategic fit.
- Exploration programs, which represent a fraction of a company's portfolio value, are generally worth less than $2/boe.
- Refiners historically generate marginal profits, if any.

Knowing the high multiples that are paid for reserves in stock-based mergers, companies should take some comfort when needed to elevate a cash bid for an attractive property by a modest percentage to entice a seller to close a deal. When the cost of reserves or the cost of production multiples is plotted for each transaction type (see Fig. 21–1A and 21–1B), it is easy to spot the premium paid in the stock-based transactions.

Company Mergers

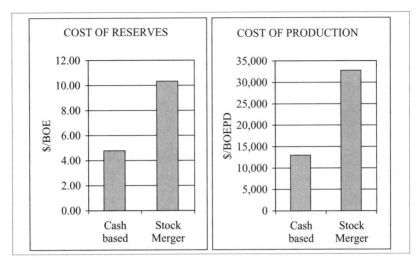

Fig. 21–1A* and *21–1B Comparison of price paid by transaction type

Companies that merge and use pooling accounting are limited in their property sales programs for a two-year period. The benefits of the favorable accounting treatment are offset in part by this restriction, and if sales are made that run afoul of pooling accounting guidelines, the acquiring company may be forced to change to the cash-based purchase cost accounting treatment. This would force the acquiring company to recharacterize the transaction, make the purchase with cash, and allocate a cash price to each acquired asset. Because of this extreme penalty, companies will obviously follow the strict guidelines that are imposed. These guidelines include, but are not limited to, the following:

- Any property sale during the two-year period immediately following closing must be consistent with the business model that would have been present in the absence of the merger, and not appear to be the result of the merger.

- The total of property sales during the same two-year period must be less than a specified percentage of the
  - Total book value of all company assets
  - Company's market capitalization
  - Company's consolidated revenues and operated income

Thus, for the period immediately following a merger that used the pooling accounting treatment, all property sales are reviewed closely to ensure compliance with these provisions.

ACCOUNTABILITY 22

Successful acquisition and divestiture programs have several common elements, one of which is to have periodic accountability analyses that compare postacquisition performance to the predicted results. An acquisition is probably scrutinized more closely and more frequently than any other expenditure in a company's budget.

The primary drivers in value determination (reserves, production rate, investments, and operating costs) are the parameters that are compared to the plan after some time has passed and the asset is absorbed into the new company. It becomes very obvious with little evaluation effort where the problem is if the performance has fallen short of expectations. The good news is that the deficiencies will be remembered and acted upon to improve the profitability of the next acquisition.

Figure 22–1 provides a simple template that can be used to track the volume and cost data for all transactions. For acquisitions, all of the data is tracked and the preacquisition evaluation is compared to actual performance for as long as management wants to conduct a review. For divestitures, the data are more limited in what can be tracked because the actual costs that are incurred by the buyer will not be known.

Company: ABC Production Inc.   Effective Date: July 1, 2001

Transaction description: Acquire 40% GWI in the Bonanza Field for $5,400,000.

Forecasted in Original Economic Analysis

Year	Production BBL	Production MCF	Product Pricing $/BBL	Product Pricing $/MCF	Revenue $M	OpCost $M	BIT Rev $M	Investment $M
2001	200000	100000	19.00	2.60	4060	650	3410	0
2002	160000	80000	19.57	2.68	3345	670	2676	600
2003	128000	64000	20.16	2.76	2757	690	2067	0
2004	102400	51200	20.76	2.84	2271	710	1561	0
2005	81920	40960	21.38	2.93	1872	732	1140	350
2006	65536	32768	22.03	3.01	1542	754	789	0
2007	52429	26214	22.69	3.10	1271	776	495	0
2008								900
Totals	790285	395142			17118	4981	12138	950
	856	Mboe						

Benefits claimed: ABC would become operator of the field and lower lifting costs.

Post Transaction Evaluation

Year	Production BBL	Production MCF	Product Pricing $/BBL	Product Pricing $/MCF	Revenue $M	OpCost $M	BIT Rev $M	Investment $M
2001	220000	115000	21.00	3.00	4965	550	4415	0
2002	165000	92000	23.00	2.80	4053	567	3486	680
2003	120000	58000	27.00	4.10	3478	583	2894	0
2004								
2005								
2006								
2007								
2008								
Totals	505000	265000			12495	1700	10795	680

Comments and lessons learned:
Pricing has exceeded forecast, 3D seismic was not the quality that was expected.

Accountability update by: _____   Date: _____

Mboe = thousand barrels of oil equivalent

Fig. 22–1 Accountability report for asset transactions

One very important ingredient to success is the timing of investment projects. Many acquisitions result in a lower rate of return than projected if the co-owners question authorizations for expenditure (AFEs) for well work that may have been unexpected. Sometimes, partners ask for an operator meeting to inquire about the new company's overall plans before they will approve any AFEs. This invariably results in delays that were not planned by the buyer.

It is best if the company uses a consistent approach to its preacquisition analysis and the postacquisition review. The isolation of key parameters and ease with which comparisons can be made is much simpler when the processes are consistent and integrated.

LESSONS LEARNED 23

There are a number of lessons that have been learned from experience that are summarized here. Some have already been mentioned in previous chapters but are worth repeating.

Processes

- The acquisition process is complicated and full of pitfalls. Potential buyers must participate in enough attempts to maintain technical competency, yet must not pursue inappropriate candidates because of the cost.
- Corporate speed and enthusiasm are assets; a slow response or approach to any phase of the process is generally not helpful in making an acquisition.
- The screening process to select the acquisition targets that best fit company strategy should be done at a high level, quickly and efficiently, and conducted by an experienced individual.
- Good acquisition teams include individuals who have also worked in sales programs and have experience understanding the opposite perspective.

Interdisciplinary Team

- A thorough analysis requires input from several disciplines. The more an asset is studied, the better the risks and upside potential can be understood, resulting in an accurate assessment of value. For properties having a relatively long life and upside potential, geoscientists, geophysicists, petrophysicists, drilling and completion engineers, operations and facilities engineers, and reserve and production engineers are needed.

- The team should be focused and excited about making the acquisition. This team should also have some accountability and associated incentives relative to the performance of the property after the acquisition.

- Competition is keen; many companies do not have strong exploration programs and must therefore make acquisitions to maintain produced volumes. Successful companies form teams that have defined responsibilities and deliverables that are consistent from one acquisition to the next.

Evaluations

- An understanding of the proved reserves is critical to the determination of the value basis of the property. Identification of upside potential is critical to the buyer to differentiate its offer from the competition. Similarly, the seller needs to value the upside to judge whether a sale is preferable to retention of the property.

- After the production forecast is finalized, it should be plotted with the prior three years of production to show any deviation from the established decline curve. Anomalies that appear should have a good explanation.
- It is wise to devote the necessary effort to performing a complete study to properly assess asset value. Preacquisition mapping to understand the geology is very important and is needed to avoid major mistakes in the evaluation.
- The seller should show the upside potential in an unrisked fashion, as risk perception varies by buyer and there only needs to be one buyer who places minimal risk on the potential for a high offer to be made.
- A thorough review of abandonment and environmental obligations is essential, even if the costs are to be spent in the distant future.

Negotiations

- Ensure that the lead negotiator has a good understanding of the value and cost drivers of the transaction, and that the evaluations engineer has a good understanding of the terms and conditions in the agreement and knows what is meaningful versus what can be negotiated away with little-to-no cost. Team members should work closely together to negotiate the best result, both in price and terms.
- Don't have multiple negotiators working various issues, as this leads to confusion and the other company adopting a divide and conquer approach. If it is necessary to have more than one negotiator, ensure that

good communication exists between them during all phases of the process.

- Provide your list of deal killers in the beginning of the negotiations to avoid giving a negative surprise after the negotiations are well under way. Stating a deal killer later in the process when it could have been identified earlier is a sure way to tarnish your company's reputation.

- Before giving in to anything, attempt to obtain the other party's whole wish list so you know the entirety of what they want and how important it is to them. This allows you also to assess the relative value of each item before responding. Avoid giving out your wish list as long as possible to maintain flexibility. After learning the extent of what the other party wants, you can then disclose your list in a more knowledgeable manner.

- Determine what the other party values and it's relative currency worth. Be prepared to give up on points or issues that you may have wanted but that have a higher currency value to the other party.

- Never discuss the merits or value of a new issue or question in front of the other party when it is raised the first time in his or her presence. Even if of no apparent internal value, it may have currency value to the other party that can be exploited. Indicate a need to discuss it with others to be able to reach a decision.

- To agree to something and then have to advise the other party later that a mistake was made and that a higher authority must also agree to it is not a positive step in negotiations. Know what terms you can agree to and what terms require approval from a higher level at the outset.

- If the other party indicates he or she can approve something when it is first mentioned, it may be a sign that you can negotiate harder on that point to get him or her to go to a higher authority level.

- If at some stage, closing a deal is of nearly equal value to not closing the deal, no negotiating to your desired position on any issue should be done.
- In setting deadline dates, note bank and business holidays, and calendar versus business days when specifying a total period of time.

Execution

- Identify fields to be sold and divest them to others simultaneously at the first closing. When buying a package of assets, do not let the fields that do not fit company strategy enter your portfolio.
- Rapidly implement the drilling and well rework program along with operational improvements.
- Include the core of the acquisitions team on the field operations team for the first six months after the acquisition is made.

In summary, successful companies have a true teamwork atmosphere with experienced professionals. Figure 23–1 shows the scope of expertise that should be drafted into the team to provide good coverage.

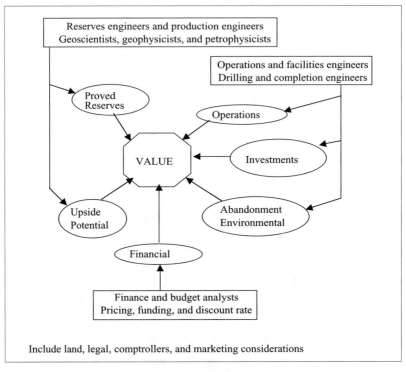

Fig. 23–1 Influence diagram with project participants

The closing recommendation is to avoid betting the farm on any acquisition regardless of its attractiveness. Given the risk and uncertainty involved in oilfield activity, there is always a chance of failure. If a company makes an acquisition that has a worst case scenario that would cripple the company, statistically it will happen at sometime to someone. Don't let it be you!

CASE HISTORIES   24

The case histories in this chapter show applications of the guidelines suggested in the book in actual purchase/sale transactions. The following three categories of deals, which are illustrated in the case histories, include the majority of asset transfers that are made in the industry:

Category 1: The purchase and sale of quality producing property (7 examples)

Category 2: The assignment of producing property in its declining years (2 examples)

Category 3: The trading of property (3 examples)

These case histories also illustrate that no two transactions are identical. There is no cookie cutter that can be used to proceed through the negotiating process. Only training, good judgment, and experience will successfully guide the team through the many decision points on the road to a profitable transaction.

Gulf of Mexico: Exploration Well Purchase/Sale
(Category 1 Deal)

A thick Miocene gas zone was logged and cored in an exploratory well that was drilled from an existing platform. Although a high rate completion was predicted from log and core analysis, the data available for determining the reservoir size were inconclusive.

The nonoperator with ownership in this discovery was reluctant to make the investment that was needed to complete the well because of the reserve uncertainty. The present value of the project was noncompetitive in the portfolio when compared to other unfunded opportunities. Given the technical hurdles and the reluctance of their partner, the operator would not commit to a plan for development. As a result, the nonoperator proceeded alone in a marketing effort to sell the field, including the new well. The operator had a preferential right to purchase.

A consultant's engineering report was not obtained to assist in the marketing effort, as the calculated proved reserve volume would have been quite low in comparison to the potential upside that the seller wanted to show. An in-house presentation was prepared to emphasize the upside potential.

The property was shown to eight companies. Only one offer was received, and it was disappointing in comparison to the upside value of the well. The market value was then recalculated using a decision tree analysis that gave appropriate weight to each development option. Using this evaluation, a more reasonable in-house market value was determined and the offer was accepted. It was clear that most of the companies who had looked at the offering did not believe the project would return an acceptable rate of return using a risked analysis.

After the deal was closed, the buyer obtained control of the project by acquiring the balance of the interest in the well from the operator. This enabled the buyer to halve the overhead that was devoted to the property by doubling the reserves that were acquired.

The buyer completed the well within three months after closing the deal. The reservoir produced for several years and ultimately recovered the majority of the upside potential.

Participant objectives

Buyer: The medium size independent that made this acquisition met a number of objectives in its acquisitions program with this purchase. Reserves were acquired that were in the early stages of the life cycle, thus there was no exploration risk. This buyer also has a technical staff with an outstanding reputation, and was confident of the reserve volume that was calculated despite the challenging lack of technical data. The infrastructure of the platform and pipelines was already in place, thus this cost and construction delay was not incurred. The company also wanted to acquire wells that would have meaningful impact to its financial results in a hurry. The high producing rate of this well, combined with the acquisition of the operated working interest, guaranteed a healthy stream of cash flow. The operator was also a company that the buyer had made acquisitions from in the past, and used this relationship to eliminate the potential loss of the sale from a preferential right election.

Seller: The major that divested the property met its goals with the sale also. The projected reserves and revenue from the well placed the field at the bottom of the company's producing property portfolio. If the company had completed the well and then sold it because of its low impact, investment funds that could have been better spent on an alternative project would have been lost from the budget. The lack of reservoir knowledge, absence of operator commitment, and lack of control also added unacceptable uncertainty to the project.

Lessons learned

The buyer was aggressive, technically capable, and used an acquisitions process that had been successful in the past—all the ingredients of an experienced approach.

The seller wished that the price had been higher, but was not willing to accept the risks associated with project participation. The fair market value was a discounted price that reflected industry's perception of the project at the time.

South Louisiana: Acquisition Proposal for Old Inland Field
(Category 1 Deal)

A startup company approached a major with an aggressive offer to acquire an old inland field that had more than 100 wells and numerous production facilities. Only six producing wells remained and the field life was projected to be less than five years. Several sidetrack and rework opportunities were identified that the major would not fund because of budget constraints. The field value was dropping at the rate of the accelerating decline curve, so the major decided it would consider divesting the property.

The prospective buyer then informed the major that the offer was contingent on an engineering report in which the whole field was to be studied. The buyer indicated the offer was "firm," but needed the engineering report to obtain financing and validate the public data that were the basis for the offer. The buyer also suggested the report could be mutually beneficial. If the report identified upside, the purchase price would be increased.

Because of the limited financial strength of the buyer, a substantial bond was required to insure that the abandonment obligations would be satisfied. The major did not want these responsibilities to revert back to it by state decree at some time in the future because of nonperformance.

The major was not pleased that the buyer needed an engineering report to submit a firm offer. In the major's view, the offer was simply a tactic to open the door to initiate the process, lock the field up for an exclusive negotiation, and ultimately reduce the offer after the engineering report was finalized.

The buyer was not pleased at the major's requirement for a costly bond to guarantee performance of abandonment. However, the buyer understood that all sales of this type by majors are normally accompanied by some assurance that the abandonment obligations will be fulfilled.

Both parties wanted to make the deal work, and agreed to continue the process and negotiate toward a closed transaction.

The buyer then provided an exhaustive wish list for information to the major to begin the study. The major realized that they were being asked to create the equivalent of a data room for the buyer's engineering consultant. The major advised if this path was followed, once the data was gathered it would only be prudent to have a data room and invite other companies to bid. The buyer was told to call its bid firm at this point and close the deal at that price, or to allow the field to go to a competitive bidding process. The buyer chose to disengage and cease communicating with the major, rather than lose exclusivity and bid against other companies.

About this time, gas prices began their unprecedented run-up to $9/Mcf in the winter of 2000. With net revenues quadrupling and the daunting prospect of a data room on the horizon, the decision to market the field was reconsidered. Investment funds were allocated to the better opportunities in the field with profitable results. Field production and cash flow increased quickly, with an associated decrease in unit lifting costs. The field was returned to core status and would not be a sales candidate for several years.

Several months later, the potential buyer contracted with a broker to sell the limited assets his business had accumulated. Had the above referenced sale occurred, the abandonment protection would have been invaluable, as the buyer of the assets was a small company with a low net worth relative to the abandonment exposure.

Participant objectives

Buyer: This small company wanted to increase its size and saw the purchase of this field as an avenue to achieve that goal. The acquisition would have more than doubled the company's producing volume and added numerous investment opportunities to the portfolio. By approaching the major with a high offer, the lowball reaction was avoided successfully. The major was almost hooked into negotiating a sale without industry competition, were it not for the request for an independent reserves study.

Seller: This proposal would have given the major a chance to monetize the marginal investments that were projected to go unfunded. Abandonment protection in the form of a bond would have been mandatory if the transaction had proceeded. The need to open a data room to market the field to industry would have been the only way to establish the field's competitive market value.

Lessons learned

The buyer almost met his objectives, and may have done so if the engineering study had not been necessary. If the buyer had been in a position to study the few reservoirs with the highest value, it might have salvaged the transaction. A larger independent would have had the resources to take the preoffer evaluation to a higher level of accuracy such that the offer would not have had to be qualified. The buyer decided to sell the company and capture the profits from the gas price increase during the period, rather than to increase company resources for expansion.

The seller was forced to study this field more thoroughly to better understand the hidden value. This proved to be a blessing as the investment opportunities were ready for funding when the decision to retain the field (rather than market it) was made.

South and North Louisiana: Purchase/Sale for a Nonoperated Package
(CATEGORY 1 DEAL)

The majors had thousands of low-working interest, nonoperated, producing properties in Louisiana from the 1950s through the early 1990s. Most of these fields were in a *caretaker* status and other than the occasional rubber-stamped AFE, received little attention. A startup company viewed this situation as an opportunity to build a diversified portfolio at moderate cost.

The startup contacted one of the majors, presented a proposal, and described the company business plan. The major was asked to select a package of approximately 30 low-working interest fields. The data needed to fully evaluate the reserves and upside potential were not available because of the limited emphasis that was historically placed on the fields by the major. As a result, a purchase price was proposed that was based upon market multiples ($/boe, $/boepd, and number of years cash flow), modified by the risked value of old gas imbalances and the projected abandonment costs.

After some negotiation on field selection, price determination, and bonding requirements, the transaction was closed. The package proved to be a mix of opportunity as was suspected. Several preferential rights were exercised, a number of profitable locations were drilled, some fields were abandoned in the near term, and a few of the gas imbalances were not recoverable.

The primary risks for the major were:

- Selling the fields without a reserve evaluation
- Being unable to quantify the upside potential

The primary risks for the startup were:

- The daunting task of absorbing the files and data while creating company contacts
- Getting *up to speed* with the downside risks of the fields
- The potential for unexpectedly high abandonment obligations
- The probability that co-owners of the better fields (which had preferential rights) would exercise those rights, and that the fields would be lost from the package
- The lack of a deterministic engineering evaluation giving a true value for the properties

Participant objectives

Buyer: The purchaser benefited in several respects with this acquisition:

- It gained immediate access to a large portfolio of producing property.
- The operating infrastructure already existed for all of the fields.

- Statistically there was the likelihood that upside did exist on some of the properties.
- New relationships and spin-off activities with the operators and co-owners helped to build the new company.

Seller: The major also benefited from the transaction in several respects; it:
- Monetized a large number of fields that had no impact on company metrics.
- Settled numerous small gas imbalance disputes that had festered for years.
- Divested more than 300 property codes from the accounting system.
- Allowed the small technical group that was responsible for nonoperated property to focus on the higher valued assets.
- Eliminated ties to the *mom and pop* operating companies who skirt regulations in ways that could negatively impact the major if a problem in field operations occurred.

Lessons learned

The primary lesson of this transaction is that buyers are resourceful and make proposals that can break down traditional methods of analysis. In this case, both parties were very satisfied with the outcome.

Gulf of Mexico: Purchase/Sale of Sunset Properties
(CATEGORY 1 DEAL)

A major was approached by the producing subsidiary of a global services company that performed platform abandonment as one of its core businesses. The subsidiary wanted to purchase a field that had just two years of positive value left, followed by several years of minimal production before depletion.

Negotiations to establish the purchase price rapidly closed in on a narrow value band from which neither company would budge. The buyer emphasized that with only two remaining producers, there was a high production risk. If just one well developed a mechanical problem, it would cause a great loss of value and the deal would be unprofitable. The seller emphasized that a higher offer was justified by the potential upside in pricing and the likelihood that mechanical problems would not develop.

The impasse was broken when the buyer asked if the major had any platforms that would need to be abandoned in the near term. The major had a platform in the Western Gulf of Mexico that needed to be salvaged that year. The buyer's parent company added a competitive cost to abandon the platform to the purchase offer of the subsidiary for the producing field. Taken together, the transaction was acceptable to the major and the deal was closed.

Results and objectives

Buyer: The buyer wanted to build a portfolio at minimal cost while keeping within the financial parameters dictated by the parent company. The wells did not develop mechanical problems, and one well produced at low pressure longer than forecasted, resulting in higher-than-expected recovery and profit.

Seller: The seller wanted to divest the low-value field and platform abandonment responsibility. The first field was divested at a fair, risk-adjusted price, and the second was abandoned at a competitive market rate.

Lessons learned

The primary lesson to be taken from this transaction is that when two companies have reasonable objectives and demands, continued communication and effort to reach common ground *after* the deal may seem undoable will normally result in success for both parties.

South Louisiana: Purchase/Sale With Proactive Co-Owner
(Category 1 Deal)

A major and a large independent operated adjacent leases in a marsh field for several decades. Synergies of operations were always obvious between the two companies, but no mutually beneficial projects were ever proposed. Extensive reservoir overlap caused each company to competitively operate a number of units that were in close proximity to each other.

As the development opportunities declined and the field slid into its sunset years, the major did not initiate a plugging program on its wells. The value of its properties began to be severely impacted by the looming abandonment costs. With the falling field value came even less attention to well work, which did not go unnoticed by the independent.

The company contacted the major and asked if an unsolicited offer to purchase would be considered. The major encouraged the submittal of an offer because it had many reasons to support a negotiated sale. This independent:

- Was the natural buyer because of field knowledge and unit participation
- Would require a fraction of the due diligence compared to any other company
- Was relatively large, financially responsible, and could be counted upon to perform the abandonment obligations
- Had a reputation as a company that paid top price for acquisitions

The major's only good reason to tell the independent that the field was not for sale was that a field of this size might attract a higher offer from an outside party if it was exposed to the market. In spite of this, the major advised the company to submit the offer. The offer was far in excess of the calculated field value and expectations. After minor negotiation, a deal was struck and the transaction was closed.

Results and objectives

Buyer: The buyer was an aggressive operator and was willing to fund the investment opportunities that were identified and ready for funding. Combined with the operating synergies and removal of marginally economic lease-line competitive activity, the purchase was very profitable for the independent.

Seller: The major felt that the sale price was more than fair because the field value would have continued to drop as the passive operating mode was continued.

Lessons learned

Both companies were very pleased with the outcome of this transaction, which shows that to be a successful buyer it is helpful to maintain relationships with your competitors. When the door of opportunity opens, take advantage of it.

East Texas: Purchase/Sale of Property With Engineering Uncertainty
(CATEGORY 1 DEAL)

A broker approached a major with an unsolicited offer to purchase a small nonoperated working interest in one of Texas's largest fields. The field had produced oil for more than 50 years with a huge overlying gas cap that was shut in since discovery. The offer was enticing, as it represented nearly five years of net cash flow from oil rim production. However, when the potential value of the gas cap was considered, the offer appeared low.

Since the major had no compelling reason to sell the field, the offer was rejected. Three more times the broker returned with a sweetened offer. Ultimately the major advised the broker that the offer was high enough that a field evaluation would be made and a sale would be considered.

The evaluation indicated that the oil reserves could be quantified with reasonable certainty, given the long producing history and limited potential for further exploitation drilling into the oil rim. The only uncertainty that was troubling revolved around the gas cap value as noted by the following concerns:

- Did the gas cap volume grow in size over the years from loss of reservoir pressure?
- Were reserves lost to an increase in the residual gas saturation?
- Was the cap partially depleted because of high gas–oil ratio production from updip oil wells?
- Would updip development drilling into the cap show more faults than are currently mapped?
- What is the current reservoir pressure in the gas cap?
- When would the operator choose to blow down the cap, and would it be produced slowly to maximize liquid recovery or quickly to maximize the present value?

It was impossible to answer all of these questions with certainty, so the major assumed the optimistic case for each and calculated a high-side value. Using this approach, a high counteroffer that ensured field value was captured regardless of the reservoir conditions or the field depletion plan was sent to the broker. This offer was accepted by the buyer.

Results and objectives

Buyer: The purchaser accepted the counteroffer and proceeded to close the transaction. Ultimately the interest was resold several years later after the gas cap blow down was initiated.

Seller: It was a good sale for the major because the operator continued to produce oil from the field in a passive manner. Additional value from the field was not created by accelerating the gas cap blow down.

Lessons learned

This case showed that an aggressive approach on behalf of the buyer was rewarded, although the price paid was very high. Unfortunately, the sale demoralized the employees of the major because there was no need to sell the field and the sales proceeds were not redeployed to acquire additional producing property in that operating area.

Gulf of Mexico: Exploitation Strategy Purchase/Sale
(Category 1 Deal)

A major owned a nonoperated interest in a large complex of several blocks that produced to a common platform for processing. Seismic data were available that the operator had not purchased. The major identified upside potential from the seismic and wanted to pursue an exploitation program. The major wanted to gain field operations as well as acquire various smaller partner interests to increase field ownership and simplify field administration prior to sharing the information with the co-owners.

The major sent offers to all of the partners and ultimately negotiated a purchase from five of them. A premium price was paid for the proved reserves because there was some expectation of improved recovery. The operator would not consider divestment and indicated that operations would not be given up easily.

The major presented to the partners the drilling plans and the benefits of transferring control of the field from the current operator.

The partners voted for the major to become the new operator on the strength of the presentation. Once operations transitioned to the major, AFEs were submitted to the partners for the drilling program.

Much to the major's surprise, the partners requested technical meetings to critically review the well proposals, creating a significant delay to the desired drilling timeline. After several meetings and a loss of three months, approval to drill was obtained.

Results and objectives

Buyer: The major considered the acquisition effort to be successful as several co-owner interests were purchased and operations were obtained.

Seller: The sellers received a good price for the risk-adjusted reserves. In each case they had not purchased the seismic data and were not cognizant of the upside potential.

Lessons learned

Several lessons are apparent from this transaction:

- Operators must stay focused and take their duty to exploit fields seriously, as partners will always be ready to capitalize on a weak effort.
- One cannot competitively value a property if others have more data, particularly data that could have been obtained.
- Good technical work is rewarded.
- It may be optimistic for a new operator to assume that drilling programs will be approved quickly by partners.

Gulf of Mexico: Gas Price Impact on Assignment of Interest
(Category 2 Deal)

Several majors owned a sprawling offshore field that covers several blocks with diverse ownership in each. One major's primary ownership was in two blocks that were in an advanced stage of decline compared to the other, more profitable blocks. This leasehold was placed in a package with several other offshore, nonoperated properties of similar value and marketed to a wide range of companies. After several months of marketing, the package did not sell. Low-value, nonoperated offshore property is not an attractive asset. It was decided to contact each of the operators in the package and attempt to negotiate individual deals for each field.

The operator of the field was contacted. There was minimal interest because the blocks had limited-to-negative value. It was summertime and the gas price was relatively low. The major was asked to pay the operator to take the field off their hands, because the low-side evaluation indicated negative value. The major declined the counteroffer and held the property.

Both companies sidelined the deal as this transaction was not a high priority for either one. After several months passed, the major contacted the operator and made the following points:

- Moving into the winter, the production volume was holding steady while the gas price was more robust and expected to remain high for several months.
- If the effective date was held to the original date, six months of interim revenue could be paid to the operator at closing.

The operator accepted this proposal and the transaction closed.

Results and objectives

Buyer: The operator wanted to acquire the interest because the field synergy with the surrounding blocks created cost savings. When the operator eventually would sell the field complex in the future, a higher working interest would mean higher buyer attraction.

Seller: The major accomplished the objective of divesting the field from the portfolio without having to pay a company to take it. It had no redeeming characteristics and an assignment was acceptable.

Lessons learned

¤   Low-value, nonoperated property is very difficult to divest.

¤   Attention to pricing cycles can be helpful in the divestment of these marginal assets.

Gulf of Mexico: Tax Impact From Assignment of Interest
(CATEGORY 2 DEAL)

A major owned a nonoperated working interest in a lease that was acquired in the mid-1980s for a bid that exceeded $100 million. A discovery was made that was thought to be significant and a high volume of reserves was booked.

Within a few years, it was apparent that the reserves were seriously overstated. The reserves were not written down because of the negative earnings impact that would have been generated from the high lease cost. As the producing life of the field approached its sunset years, the remaining tax basis was also huge relative to the actual value of the reserves. An attempt was made to sell the field as it declined into marginal value, but no bids were received. The decision was made to hold the field until depletion and handle the write-offs at that time.

Upon reviewing the evaluation for a second time, it was noted that the tax basis for the lease was omitted from the analysis. When the huge tax basis was included, the major realized that more near-term, after-tax cash flow would be realized with a much higher present value if the property was simply assigned to the operator, so the tax basis could be written off on the tax return as a surrender of property.

The operator accepted the assignment while the field still had some positive value remaining, and the major wrote off millions of remaining tax basis as a tax deduction.

Results and objectives

Buyer: The operator in this example did not make an offer while the property was marketed, knowing that its chance to acquire the asset would come with a preferential rights election if the field was sold. Ultimately, it was not sold but the patience paid off when the seller approached the operator to take the field on assignment.

Seller: The goal of the seller to divest the property was met, and the write-off of the tax basis resulted in a high tax deduction. Valued on a present-worth basis, the deduction was worth much more taken at the present time than if it was taken at depletion.

Lessons learned

An unusual tax basis for a property (that is very high or very low relative to property value) will influence the evaluation and the ultimate decision to dispose of a property potentially more than the present value of the reserves.

Gulf of Mexico: Trade of Property, Evaluation Expertise
(Category 3 Deal)

Two majors reviewed their portfolios and noted serious differences in the perception of upside and future investment potential on four fields that were co-owned and in the mature period of their life cycles.

Field A had high abandonment cost, minimal production, and undrilled potential. The operator (Major 1) was pessimistic toward the exploration and would not drill. The nonoperator (Major 2) liked the potential and wanted to acquire the field.

Field B was operated by Major 2 and was situated in the midst of a platform complex operated by Major 1. The field had strong cash flow and low-risk exploitation potential. Major 1 wanted to acquire this field for the upside and synergy savings.

Field C was an old gas unit that was projected to deplete within 6 months. Major 1 was overproduced by nearly four billion cubic feet (BCF) and did not want to settle the gas imbalance in cash for an estimated $6 million.

Field D was operated by Major 2, had moderate worth, and was not strategic to either company. It was viewed as a value equalizer for the transaction.

After several months of negotiation in which other co-owned properties and alternate currencies were considered, a trade was struck for these four assets; Major 1 divested Fields A, C, and D to Major 2, and Major 1 acquired Field B from Major 2.

Results and objectives

Major 1: This company got the better deal. Field B provided two exploitation locations, has produced as expected, and enabled the company to lower lifting costs. Also, they avoided payment of $6 million in cash for the gas imbalance in Field C.

Major 2: After the geophysical evaluation for Field A was completed, it was decided that the prospects were not viable and they were not drilled. Thus, the value of the field plummeted from $6 million to <$3 million> because of the pending abandonment. Field D did produce as expected for several years prior to depletion.

Lessons learned

- The company with the better technical staff will nearly always generate a more accurate forecast of field performance.

- The consideration of alternate currencies, in this case the inclusion of the gas imbalance, was a smart tactical move.

South Louisiana: Innovative Approach to Property Trade
(CATEGORY 3 DEAL)

A major owned the majority of a large, operated gas field that had upside potential. The co-owner had a small interest through an old unit agreement. The co-owner did not have seismic data coverage for the bulk of the reservoir area that was leased by the major.

The major performed an analysis of the property and made a cash offer to the co-owner, which was rejected as being too low. Subsequent discussion indicated that in reality the co-owner did not know the value of the field because it did not have sufficient information to determine the value of the property. The safe path for them was to simply reject the offer rather than divesting at a price that was too low.

The major wanted to own the property in its entirety, as it had an aggressive well work and development drilling program planned. If acquired, ownership of the 100% working interest would provide total operational control and a relief from the AFE process.

The major then decided to approach the acquisition as an exchange in which the co-owner could keep a share of the upside potential and eliminate price volatility from the analysis. The next proposal was to acquire the working interest in exchange for an equally valued overriding royalty interest. Both companies calculated the royalty that approximated the value of the working interest, and subsequently negotiated to a 4.4% royalty from the whole unit.

Results and objectives

Major: The major was able to perform the well work and drill the upside potential on an accelerated timeline, as well as capture infrastructure operating cost synergies in adjacent fields that were already owned 100%.

Co-owner: This company was pleased to benefit from the upside potential by virtue of the overriding royalty interest, particularly when, as a working interest owner, it would have been challenging to spend its share of the upcoming investment with limited information.

Lessons learned

Flexibility in keeping focused on the objective was crucial for the major in this transaction to avoid getting discouraged when the co-owner balked at divesting its interest. At some point in the future— when it begins to impact the economic limit and ultimate reserve recovery from the property—the major will approach the co-owner again to purchase the royalty interest.

Gulf of Mexico: Trade of Property for Consolidating Efficiencies
(CATEGORY 3 DEAL)

Two majors each had interest in two different properties that had no other partners.

Major 1 owned more than 95% of a huge producing asset and wanted to acquire the balance. This would provide operational control, simplify accounting, and enable a prospective exploitation program to proceed without partners. There was a chance also that a deeper exploration objective may have been overlooked on the seismic data that was originally shot over the prospect.

Major 2 owned and operated an asset that was much smaller on a 50-50 basis with Major 1. Major 2 had identified a few exploitation locations that Major 1 would not fund. Acquisition of the field would allow a drilling program to proceed.

After a short negotiation, the companies agreed to swap the properties and settle minor gas imbalances in the process.

Results and objectives

Major 1: This company was pleased to acquire the balance of the large field. Aside from the operational and accounting simplification, the field has a long remaining life and upside potential continues to be identified. Also, the facilities have become an unexpected revenue source by processing off-lease production.

Major 2: This company was also pleased to acquire the balance of its operated field, as the drilling program was able to proceed and the benefits of the geological and geophysical expertise were captured at 100% of property value rather than at only 50%.

Lessons learned

This transaction is one that can be duplicated in many other cases. However, it is becoming increasingly rare for majors to own a portfolio that is large enough to have the scope of overlap needed to identify fields that are such a good fit and thus enable trades to be made. Many independents own sizable portfolios in the Gulf of Mexico now, and trades of assets between independents may be more common in the future.

The large field acquired by Major 1 has had two revitalization programs. The old adage that 'big fields get bigger' was certainly true in this case, and ultimately it will be shown that Major 1 got the better of the trade because the field they acquired has a much longer life with relatively high upside potential.

INDEX

3-D seismic data, 50

A

Abandonment obligation/operations, 24, 69, 99, 117–118, 130, 153

Accountability report, 147–148

Accountability, 147–149:
  accountability report, 147–148

Accounting records review (due diligence), 125–127

Acquisition and divestiture (successful company profiles), 1–10:
  players, 1–10;
  buyers, 1–5;
  sellers, 5–9;
  business relationships, 10

Acquisition competition, 60–61

Acquisition evaluation, 135–138

Acquisition prices (historical), 135–138

Acquisition process, 5–9, 13–15, 55–64:
  steps to the process, 55–57;
  proactive and reactive approaches, 57–59;
  data room visit, 59–60;
  reserves assessments, 60–61;
  financing options, 61–62;
  product price considerations, 62–64

Acquisition proposal for old inland field (South Louisiana), 160–163:
  case history, 160–163;
  participant objectives, 162;
  lessons learned, 162–163

Acquisition target (approach), 57–59:
  proactive, 58–59;
  reactive, 59

Activity timeline (divestment process), 47–49

Allocation process (price), 114–115

Alternatives/options (constructing the offer), 99–101

Amount strategy (constructing the offer), 104–106

Appraised value (constructing the offer), 101–104

Approval (government), 129–130

Assessment of reserves (acquisition process), 60–61

Asset acquisitions (market multiples), 74–79

181

Asset sale/purchase (motivating/
    demotivating factors), 11–15,
    17–21:
    seller, 11–13, 17–19;
    buyer, 13–15, 19–21

Asset study, 17

Asset transfer categories, 157

Assignment of interest (case histories),
    173–175:
    gas price impact, 173–174;
    tax impact, 174–175

Auction proceeds, 34, 36–41

Auctions (marketing options), 34–41:
    proceeds, 34, 36–41

Authorization for expenditure (AFE),
    148–149

Average wellhead price, 74–79

B

Bank financing, 62

Base case funding, 27

Base case sensitivity analysis
    (acquisition process), 56

Bid analysis (acquisition process), 56

Bid success, 13–15

Bond performance, 118–119

Bonding protection, 99, 117–119:
    requirement, 118–119

Bonding requirement, 118–119

Book value, 29–31

Brokers (marketing options), 42–43

Btu basis, 136

Business model, 8–9

Business/company relationships, 10

Buyer motivation/Non-motivation,
    13–15, 19–21:
    indemnities required by seller, 20;
    discovery of undisclosed liabilities,
        20;
    financing questions/problems, 20;
    lack of information, 21;
    delay requested by seller, 21

Buyer retreat, 19–21:
    indemnities required by seller, 20;
    discovery of undisclosed liabilities,
        20;
    financing questions/problems, 20;
    lack of information, 21;
    delay requested by seller, 21

Buyer strategy, 139

Buyer versus seller viewpoints
    (valuation methodologies), 73–74

Buyer's evaluation (new field
    discoveries), 95–96

Buyers (successful company), 1–5

C

Capital gains tax, 18, 91

Capture exploration potential
    (acquisition process), 56

Case histories, 157–180:
    Gulf of Mexico:
        exploration well purchase/sale,
            158–160;
    South Louisiana:
        acquisition proposal for old inland
            field, 160–163;
    South and North Louisiana:
        purchase/sale for a non-operated
            package, 163–165;
    Gulf of Mexico:
        purchase/sale of sunset
            properties, 166–167;
    South Louisiana:
        purchase/sale with
            proactive co-owner, 167–169;
    East Texas:
        purchase/sale of property with
            engineering uncertainty,
            169–171;

Gulf of Mexico:
exploitation strategy purchase/sale, 171–172;
Gulf of Mexico:
gas price impact on assignment of interest, 173–174;
Gulf of Mexico:
tax impact from assignment of interest, 174–175;
Gulf of Mexico:
trade of property, evaluation expertise, 176–177;
South Louisiana:
innovative approach to property trade, 177–178;
Gulf of Mexico:
trade of property for consolidating efficiencies, 179–180

Cash flow analysis, 2–5, 20, 24, 103, 142:
discounted cash flow, 103;
free cash flow, 142

Cash versus stock-based transactions, 142–144

Closing documents, 38

Closing process, 124

Collar (hedge arrangement), 5

Company impacts (acquisition process), 56

Company mergers, 5, 141–145

Company profiles, 1–10:
players, 1–10;
buyers, 1–5;
sellers, 5–9;
relationships, 10

Company strategy, 13–14

Consent to assign provision, 129

Consolidating efficiencies (case history), 179–180

Constructing the offer, 99–112:
options and alternatives, 99–101;
appraised value, 101–104;
strategy of amount, 104–106;
sales package examples, 106–112

Co-owner solicitation (marketing to industry), 44–45

Co-owner, 44–45, 95–96, 100–101, 167–169:
solicitation, 44–45;
rights, 95–96;
interests, 100–101;
case history, 167–169

Corporate sales programs, 131–133

Cost accounting, 144

Cost analysis, 101

Counter offer, 105

D

Data package, 33

Data room requirements (divestment process), 49–51:
data evolution, 50–51

Data room visit (acquisition process), 59–60

Data tracking (accountability), 147–149

Debt service, 23

Decision, 23–26, 56, 97:
fiscal, 23–26;
acquisition, 56;
divestiture, 97

Deep drilling rights, 21

Deepwater example (new field discovery), 96–97

Delay requested by seller, 21

Delineation, 65–66:
discovery value, 65–66;
life cycle risks, 66

De-motivating factors (asset sale/purchase), 17–21:
seller, 17–19;
buyer, 19–21

Depletion, 2–3, 12

Deteriorated conditions, 20

Determining price (property), 83–90:
impact of various factors, 83–85;
multiple field packages, 85–86;
sensitivity analysis, 86–88;
product price, 88–89;
reserves growth, 90

Development (life cycle risks), 66

Development cost (new field discovery), 94–95

Discounted cash flow, 103

Discoveries (new field), 93–97:
characteristics, 93;
reasons for sale, 94–95;
buyer's evaluation, 95–96;
deepwater example, 96–97

Discovery (undisclosed liabilities), 20

Disincentive (asset sale/purchase), 17–21:
seller, 17–19;
buyer, 19–21

Distressed property, 63

Divestiture and acquisition (successful company profiles), 1–10:
players, 1–10;
buyers, 1–5;
sellers, 5–9;
business relationships, 10

Divestiture, 1–10, 21, 47–53, 97:
successful company profiles, 1–10;
process, 5–9, 47–53;
decision, 97

Divestment process, 5–9, 47–53:
activity timeline, 47–49;
data room requirements, 49–51;
third-party reserves report, 51–53

Drilling rights, 21

Due diligence, 20–21, 125–127

E

Early life field, 12–13

Earnings impact, 29–31

Earnings write-downs (seller evaluation), 29–31

East Texas: purchase/sale of property with engineering uncertainty (case history), 169–171:
participant objectives/results, 170;
lessons learned, 171

Economic analysis, 86–88, 101–104:
economic evaluation parameters, 101–104

Economic evaluation parameters (property), 101–104

Effective date (negotiating the agreement), 123–124

Elevated multiples justification (industry activity), 139–140

Engineering report, 51–53

Engineering uncertainty property (case history), 169–171

Environmental obligation, 20, 24, 117–118, 130, 153

Evaluation (by seller), 23–32:
fiscal decisions, 23–26;
retention value, 26–27;
tax consequences, 27–29;
earnings write-downs, 29–31;
financial impact, 31

Evaluation parameters (property), 101–104

Evaluation expertise (case history), 176–177

Evaluation variables, 86–88

Evaluations (lessons learned), 152–153

Exchanges (like-kind), 91–92

Execution (lessons learned), 155–156

Exit strategy, 13, 68

Exploitation strategy purchase/sale (Gulf of Mexico), 171–172:
case history, 171–172;
participant objectives/results, 172;
lessons learned, 172

Exploration (life cycle risks), 65

Exploration potential, 21, 56:
acquisition process, 56

Exploration well purchase/sale (Gulf of Mexico), 158–160:
case history, 158–160;
participant objectives, 159;
lessons learned, 160

Exposure limitation, 3–4

F

Factor impact (determining price), 83–85

Fair market value (valuation methodologies), 71–72

Federal tax, 18

Federal Trade Commission (FTC), 129

Field characteristics, 12–13

Field discoveries, 93–97:
characteristics, 93;
reasons for sale, 94–95;
buyer's evaluation, 95–96;
deepwater example, 96–97

Field operations, 81

Field retention value, 26–28

Field trip (due diligence), 125–126

Financial impact (seller evaluation), 31–32

Financing (acquisition process), 20, 61–62, 100:
questions/problems, 20;
options, 61–62

Fiscal decisions (seller evaluation), 23–26

Flush production, 66

Free cash flow, 142

Future field value, 24–26

G

Gambler's ruin, 15

Gas price impact on assignment of interest (Gulf of Mexico), 173–174:
case history, 173–174;
participant objectives/results, 174;
lessons learned, 174

Government approval, 129–130

Gulf of Mexico, 78, 96–97, 119, 129, 158–160, 166–167, 171–177, 179–180:
case histories, 158–160, 166–167, 171–177, 179–180

Gulf of Mexico: exploitation strategy purchase/sale (case history), 171–172:
participant objectives/results, 172;
lessons learned, 172

Gulf of Mexico: exploration well purchase/sale (case history), 158–160:
participant objectives, 159;
lessons learned, 160

Gulf of Mexico: gas price impact on assignment of interest (case history), 173–174:
participant objectives/results, 174;
lessons learned, 174

Gulf of Mexico: purchase/sale of sunset properties (case history), 166–167:
participant objectives/results, 166–167;
lessons learned, 167

Gulf of Mexico: tax impact from assignment of interest (case history), 174–175:
participant objectives/results, 174;
lessons learned, 174

Gulf of Mexico: trade of property for consolidating efficiencies (case history), 179–180:
participant objectives/results, 179;
lessons learned, 180

Gulf of Mexico: trade of property, evaluation expertise (case history), 176–177:
participant objectives/results, 177;
lessons learned, 177

H

Hart-Scott-Rodino Act, 129

Hedging, 3–5

Historical acquisition prices (industry activity), 135–138

Historical figures (market multiples), 79–80

Holding value, 26–27

Hurdle rates, 9

I

Impact analysis (acquisition process), 56–57:
company, 56;
other, 57

Impact factors (determining price), 83–85

Indemnification (purchase/sale agreement), 20:
indemnities required by seller, 20

Independent companies, 5, 7–8

Industry activity, 135–140:
historical acquisition prices, 135–138;
justification of elevated multiples, 139–140

Information availability (property), 21, 49–51

In–house marketing (to industry), 44

Initial public offering (IPO), 14

Innovative approach to property trade (South Louisiana), 177–178:
case history, 177–178;
participant objectives/results, 178;
lessons learned, 178

Interdisciplinary team (lessons learned), 152

Internal funding (financing), 61

Internal Revenue Service (IRS), 91–92

Investment capital, 20

J–K

Justification (elevated multiples), 139–140

L

Lessons learned, 151–156, 159, 162–163, 167, 169, 171–172, 174–175, 177–178, 180:
processes, 151;
interdisciplinary team, 152;
evaluations, 152–153;
negotiations, 153–155;
execution, 155–156;
exploration well purchase/sale, 159;
acquisition proposal for old inland field, 162–163;
purchase/sale of sunset properties, 167;
purchase/sale with proactive co-owner, 169;
purchase/sale of property with engineering uncertainty, 171;
exploitation strategy purchase/sale, 172;
gas price impact on assignment of interest, 174;
tax impact from assignment of interest, 175;
trade of property, evaluation expertise, 177;
innovative approach to property trade, 178;
trade of property for consolidating efficiencies, 180

Liabilities (discovery), 20

Life cycle risks, 65–69:
exploration, 65;
delineation, 66;
development, 66;
production, 67–68;
redevelopment, 68;
sunset period, 69

Lifting cost, 132

Like-kind exchanges, 91–92

Louisiana (case histories), 160–165, 167–169, 177–178

Louisiana sales package (example), 108–109

Low production rate, 131

Low working interest, 132

Low-confidence reserves, 136–137

M

Major companies, 5–7

Market multiples (valuation methodologies), 74–80:
asset acquisitions, 74–79;
historical figures, 79–80

Market multiples, 39–41, 74–80, 139–143:
valuation methodologies, 74–80

Market value, 24, 71–72:
valuation methodologies, 71–72

Marketing in–house (to industry), 44

Marketing options, 33–45:
auctions, 34–41;
brokers, 42–43;
in–house to industry, 44;
solicit co-owners, 45

Marketing time frame, 26

Mergers (company), 5, 79–80, 141–145

Mezzanine financing, 62

Minerals Management Service (MMS), 129

Minimum bid, 38

Mississippi sales package (example), 110–112

Monetizing assets, 23

Motivating factors, 11–15:
seller motives, 11–13;
buyer motives, 13–15

Move forward (acquisition process), 56

Multiple field packages (determining price), 85–86

Multiples justification (industry activity), 139–140

Multiples, 74–80, 139–140:
valuation methodologies, 74–80;
asset acquisitions, 74–79;
historical figures, 79–80;
justification, 139–140

N

Negotiating the agreement, 121–124, 153–155:
terms and conditions, 121–123;
effective date, 123–124;
lessons learned, 153–155

Negotiations (lessons learned), 153–155

Net proceeds, 27–28

New field discoveries, 93–97:
characteristics, 93;
reasons for sale, 94–95;
risk/risk analysis, 94–95;
buyer's evaluation, 95–96;
deepwater example, 96–97

Non-company viewpoint (property analysis), 61

Non-motivating factors, 17–21:
seller, 17–19;
buyer, 19–21

Non-operated package (case history), 163–165

Non-producing assets, 142

O

Objectives/results (participant), 159, 162, 166–168, 170, 172, 174–175, 177–179:
  exploration well purchase/sale, 159;
  acquisition proposal for old inland field, 162;
  purchase/sale of sunset properties, 166–167;
  purchase/sale with proactive co-owner, 168;
  purchase/sale of property with engineering uncertainty, 170;
  exploitation strategy purchase/sale, 172;
  gas price impact on assignment of interest, 174;
  tax impact from assignment of interest, 175;
  trade of property, evaluation expertise, 177;
  innovative approach to property trade, 178;
  trade of property for consolidating efficiencies, 179

Offer (constructing), 99–112:
  options and alternatives, 99–101;
  appraised value, 101–104;
  strategy of amount, 104–106;
  sales package examples, 106–112

Oil & Gas Asset Clearinghouse (O&GAC), 35

Old inland field (case history), 160–163

Opening offer, 104–106

Opening position, 20, 104–106:
  opening offer, 104–106

Operating cost reduction, 2

Operating synergy, 18

Operational concerns (valuation methodologies), 81

Options/alternatives (constructing the offer), 99–101

Outlier bid, 14, 106

P

Package (properties), 85–86

Participant objectives/results, 159, 162, 166–168, 170, 172, 174–175, 177–179:
  exploration well purchase/sale, 159;
  acquisition proposal for old inland field, 162;
  purchase/sale of sunset properties, 166–167;
  purchase/sale with proactive co-owner, 168;
  purchase/sale of property with engineering uncertainty, 170;
  exploitation strategy purchase/sale, 172;
  gas price impact on assignment of interest, 174;
  tax impact from assignment of interest, 175;
  trade of property, evaluation expertise, 177;
  innovative approach to property trade, 178;
  trade of property for consolidating efficiencies, 179

Payout, 5

Permian Basin, 78

Players (successful company profiles), 1–10:
  buyers, 1–5;
  sellers, 5–9;
  relationships, 10

Point forward options, 97

Pooling accounting, 141, 144–145

Portfolio adjustment program, 131

Portfolio management, 5–10, 12, 17–18, 26, 131–133:
  business relationships, 10;
  portfolio adjustment program, 131

Post-acquisition review, 149

Potential reserves, 10, 13, 18, 152–153:
  proved reserves, 152–153

Pre-acquisition analysis, 2, 149
Pre-acquisition mapping, 153
Preferential rights, 95–96, 113–115
Present value determination, 24–25
Price allocation process (property), 114–115
Price deck (reserves), 136–138
Price determination (property), 83–90:
    impact of various factors, 83–85;
    multiple field packages, 85–86;
    sensitivity analysis, 86–88;
    product price, 88–89;
    reserves growth, 90
Price multiples (property), 39–40
Price premium (property), 9
Price protection (product), 3–5, 99:
    sales contract, 3–4;
    hedging, 3–5
Prices/pricing (product), 3–5, 18–19, 62–64, 88–89, 99:
    fluctuation, 3–4;
    protection, 3–5, 99;
    appreciation potential, 18–19;
    uplift, 19;
    acquisition considerations, 62–64;
    volatility, 64;
    forecasts, 88–89;
    determining, 88–89
Prices/pricing (property), 9, 14, 39–40, 74, 83–90, 114–115, 135–138, 142–144:
    premium, 9;
    multiples, 39–40;
    sale price, 74;
    determination, 83–90;
    allocation process, 114–115;
    historical, 135–138;
    transaction type, 142–144
Prices/pricing (reserves), 136–138
Proactive and reactive approaches (acquisition process), 57–59:
    proactive, 58;
    reactive, 59

Proactive co-owner (case history), 167–169
Process steps (acquisition), 55–57:
    perform quick screen, 55;
    decide to move forward, 56;
    capture exploration potential, 56;
    analyze sensitivities to base case, 56;
    company impacts, 56;
    perform bid analysis, 56;
    review other impact items, 57
Process steps (divestment), 47–53:
    activity timeline, 47–49;
    data room requirements, 49–51;
    third-party reserves report, 51–53
Processes (lessons learned), 151
Product price, 3–5, 18–19, 62–64, 88–89, 99:
    fluctuation, 3–4;
    protection, 3–5, 99;
    appreciation potential, 18–19;
    uplift, 19;
    acquisition considerations, 62–64;
    volatility, 64;
    forecasts, 88–89;
    determining, 88–89
Production (life cycle risks), 67–68
Production multiples, 39–41
Production performance, 21, 131:
    low production rate, 131
Production replacement, 17
Profit margin, 18, 28
Profit/profitability, 12, 18–19, 28, 67, 71–81:
    profit margin, 18, 28
Property evaluation, 2, 5, 24–25, 71–81, 101–104, 176–177:
    buyers, 2;
    sellers, 5;
    property value, 24–25, 71–81;
    valuation methodologies, 71–81;
    economic parameters, 101–104;
    evaluation expertise, 176–177
Property package, 85–86

Property trade (case histories), 157, 176–180:
  evaluation expertise, 176–177;
  innovative approach, 177–178;
  consolidating efficiencies, 179–180

Property valuation methodologies, 71–81:
  fair market value, 71–72;
  seller versus buyer viewpoints, 73–74;
  market multiples—
    asset acquisitions, 74–79;
  market multiples—
    historical figures, 79–80;
  operational concerns, 81

Property value, 24–25, 71–81:
  valuation methodologies, 71–81.
  See also Property evaluation.

Protection bonding, 117–120

Proved reserves, 152–153

Purchase motives, 13–15

Purchase/sale for Non-operated package (South and North Louisiana), 163–165:
  case history, 163–165;
  participant objectives, 164–165;
  lessons learned, 165

Purchase/sale of property with engineering uncertainty (East Texas), 169–171:
  case history, 169–171;
  participant objectives/results, 170;
  lessons learned, 171

Purchase/sale of sunset properties (Gulf of Mexico), 166–167:
  case history, 166–167;
  participant objectives/results, 166–167;
  lessons learned, 167

Purchase/sale with proactive co-owner (South Louisiana), 167–169:
  case history, 167–169;
  participant objectives/results, 168–169;
  lessons learned, 169

Q

Quick screen (acquisition process), 55

R

Rate life–sales price relationships, 77–78

Rate of return, 5

Reactive and proactive approaches (acquisition process), 57–59:
  proactive, 58;
  reactive, 59

Reasons for sale (new field discoveries), 94–95

Records review (due diligence), 125–127

Recovery rate, 77–78

Redevelopment, 68, 100–101:
  life cycle risks, 68

Relationships (business), 10

Reserve adjustment factors, 101–103

Reserves assessments (acquisition process), 60–61

Reserves by Btu/price equivalent, 136

Reserves definitions, 101, 103

Reserves growth (determining price), 90

Reserves multiples, 141

Reserves potential, 10, 13, 18, 67, 152–153:
  proved reserves, 152–153

Reserves price, 136–138

Reserves report (third-party), 20, 51–53

Reserves retention, 92

Reserves risk, 101–103

Reserves, 9–10, 13, 17–18, 20, 29–31, 51–53, 60–61, 90, 92, 101–103, 136–138, 152–153:
calculation, 9;
additions, 9;
potential, 10, 13, 18, 152–153;
replacement, 17;
report, 20, 51–53;
assessments, 60–61;
growth, 90;
retention, 92;
adjustment factors, 101–103;
risk, 101–103;
definitions, 101, 103;
low confidence, 136–137;
by Btu/price equivalent, 136;
unproved, 136–137;
price, 136–138;
multiples, 141;
proved, 152–153

Reservoir data analysis, 68

Resource base depletion, 2–3

Retention value, 9, 12, 17, 26–27:
seller evaluation, 26–27

Review other impacts (acquisition process), 57

Revitalization (field), 68, 100–101:
life cycle risks, 68

Right-to-purchase provision, 95–96

Risk/risk analysis, 3, 8, 10, 65–69, 94–95, 101–103:
risk-reward sharing, 3;
risk-adjusted basis, 8;
life cycle, 65–69;
new field discovery, 94–95;
reserves, 101–103

Risk-adjusted basis, 8

Risking of reserves, 101–103

Risk-reward sharing, 3

Royalty interest, 21, 39–40, 99

S

Sale motives, 11–13

Sale reasons (new field discoveries), 94–95

Sales candidate, 23–26, 31

Sales contract, 3–4

Sales package examples (constructing the offer), 104–106

Sarbanes–Oxley Act of 2002, 31

Scotia Group, Inc., 135–136

Screen analysis (acquisition process), 55, 151

Securities and Exchange Commission (SEC), 31

Seismic data, 68

Sell incentives, 11–12

Seller data evolution, 50–51

Seller evaluation, 23–32:
fiscal decisions, 23–26;
retention value, 26–27;
tax consequences, 27–29;
earnings write-downs, 29–31;
financial impact, 31–32

Seller motivation/Non-motivation, 11–13, 17–19

Seller requests delay, 21

Seller requires indemnities, 20

Seller versus buyer viewpoints (valuation methodologies), 73–74

Sellers (successful company), 5–9

Sensitivity analysis, 56, 86–88:
base case (acquisition process), 56;
determining price, 86–88

Society of Petroleum Evaluation Engineers (SPEE), 72, 101–104:
Survey of Economic Parameters Used in Property Evaluation, 101–104

Solicitation, 12–13, 44:
co-owners, 44

South and North Louisiana: purchase/sale for a Non-operated package (case history), 163–165:
participant objectives, 164–165;
lessons learned, 165

South Louisiana: acquisition proposal for old inland field (case history), 160–163:
participant objectives, 162;
lessons learned, 162–163

South Louisiana: innovative approach to property trade (case history), 177–178:
participant objectives/results, 178;
lessons learned, 178

South Louisiana: purchase/sale with proactive co-owner (case history), 167–169:
participant objectives/results, 168–169;
lessons learned, 169

State ownership, 130

Stock–based versus cash transactions, 142–144

Strategic value, 72

Strategy of amount (constructing the offer), 104–106

Successful company profiles, 1–10:
players, 1–10;
buyers, 1–5;
sellers, 5–9;
relationships, 10

Sunset property, 12–13, 24, 69, 157, 166–167:
life cycle risks, 69;
case history, 166–167

T

Tax basis, 18

Tax consequences, 18, 27–29, 31, 91–92, 124, 174–175:
tax basis, 18;
tax cost, 18, 27–29, 31;
capital gains tax, 18, 91;
seller evaluation, 27–29;
tax issues, 91–92;
assignment of interest, 174–175

Tax cost, 18, 27–29, 31

Tax impact from assignment of interest (Gulf of Mexico), 174–175:
case history, 174–175;
participant objectives/results, 175;
lessons learned, 175

Tax impact, 18, 27–29, 31, 91–92, 124, 174–175:
tax basis, 18;
tax cost, 18, 27–29, 31;
capital gains tax, 18, 91;
seller evaluation, 27–29;
tax issues, 91–92;
assignment of interest, 174–175

Technology advances, 136–137

Terms and conditions (negotiating the agreement), 121–123

Texas (case history), 169–171

Texas sales package (example), 106–108

Third-party reserves report (divestment process), 51–53

Three–tier hierarchy (company size), 5

Timeline (divestment process), 47–49

Trade of property for consolidating efficiencies (Gulf of Mexico), 179–180:
case history, 179–180;
participant objectives/results, 179;
lessons learned, 180

Trade of property, evaluation expertise (Gulf of Mexico), 176–177:
  case history, 176–177;
  participant objectives/results, 177;
  lessons learned, 177

Trade of property, innovative approach (case history), 177–178:
  participant objectives/results, 178;
  lessons learned, 178

Trade of property, 157, 176–180:
  case histories, 176–180;
  evaluation expertise, 176–177;
  innovative approach, 177–178;
  consolidating efficiencies, 179–180

Transaction delay, 21

Transaction multiples, 141

Transaction size, 79–80, 137–138:
  historical, 79–80

Transaction structures, 3

Transaction type (price), 142–144

Transactions impairing asset, 21

U

Uncertainty of property (case history), 169–171

Undisclosed liabilities (discovery), 20

Unique life cycle risks, 65–69:
  exploration, 65;
  delineation, 66;
  development, 66;
  production, 67–68;
  redevelopment, 68;
  sunset period, 69

Unit of production (UOP), 30, 132:
  write-off rate, 132

Unproved reserves, 136–137

Unsolicited offer, 12

Upside potential, 10, 13, 18, 67, 152–153

V

Valuation methodologies, 71–81:
  fair market value, 71–72;
  seller versus buyer viewpoints, 73–74;
  market multiples—asset acquisitions, 74–79;
  market multiples—historical figures, 79–80;
  operational concerns, 81

Value creation, 1, 8

Value gain, 19

Value retention (seller evaluation), 26–27

Viewpoints (seller versus buyer valuation), 73–74

Visit data room (acquisition process), 59–60

Volume growth, 2–3, 14

Volume–cost data accountability, 147–149

W–Z

Working interest, 39, 66, 90, 132

Write-downs (earnings), 29–31

Write-off rate, 132